'Every man and woman should read
Rex is an important, yet wickedly wit
which touches on the current debates
everything on its head. Pressingly contemporary, it's the ideal companion read to sit alongside *The Handmaid's Tale* and *The Power.*'

Judges, 2017 Royal Society Insight Investment Science Book Prize

'Graced with precisely focused humour, the author makes a good case that men and women are far more alike than many would claim. Feminist? Possibly. Humanist? Certainly. A compellingly good read.'

Professor Richard Fortey

'A densely packed, spirited book, with an unusual combination of academic rigour and readability … The expression "essential reading for everyone" is usually untrue as well as a cliché, but if there were a book deserving of that description this might just be it.'

Antonia Macaro, *Financial Times*

'In addition to being hopeful, Fine is also angry. We should all be angry. *Testosterone Rex* is a debunking rumble that ought to inspire a roar.'

Sarah Ditum, *Guardian*

'Packed with convincing evidence and astonishing facts, all of which seem so important that everybody should be made to read all of it immediately, or at least before typing another word on Twitter about political correctness gone mad.'

Katy Guest, *The Pool*

'Such tireless energy is just what's needed to slay a monster of this gargantuan size and this determined tenacity … Fine's funny, spiky book gives reason to hope that we've heard Testosterone Rex's last roar.'

Annie Murphy Paul, *New York Times Book Review*

'Fine knocks it out of the park with her smart and eye-opening *Testosterone Rex*'

Adrian Laing, editor of *The Amazon Book Review*

'Goodbye beliefs in sex differences disguised as evolutionary facts. Welcome the dragon slayer: Cordelia Fine wittily but meticulously lays bare the irrational arguments that we use to justify gender politics.'

Uta Frith, Emeritus Professor of Cognitive Development, University College, London

'The delusion that there are distinct and unique male and female natures, put in place by an unholy alliance of genes, hormones and neurones, remains alive and well. Cordelia Fine dismantles this myth with style, wit and scientific precision. This combination of scientific responsibility and general accessibility is desperately needed if we are to escape the serious social damage caused by such widely disseminated pseudoscience.'

John Dupré, Professor and Director of Egenis, Centre for the Study of Life Sciences, University of Exeter

'A witty, authoritative guide to how pretty much everything you think you know about gender is backwards.'

Caroline Criado-Perez, OBE, author of *Do it Like a Woman*

'There have been plenty of books about gender and stereotyping and unconscious bias. What's original in this book is that she takes apart the science so forensically.'

Claudia Hammond, presenter of BBC Radio 4's *All in the Mind*

'A fascinating, greatly contemplative discussion of sex and gender and the embedded societal expectations of both.'

Kirkus Reviews

'Fine's entertaining and thoughtful book is a valuable addition to the discussion about gender.'

Ian Critchley, *The Sunday Times*

'A provocative and often fascinating book. Armed with an array of studies on everything from rats to humans, she shows that adaptive traits can take different forms depending on the circumstances, and nothing is fixed.'

The Economist

'Cordelia Fine combines formidable intellect, forensic analysis and devastating wit to expose those myths of sex, gender and human behaviour that might just reflect testosterone-fuelled, wishful thinking. This book makes crucial reading for those with an interest, from any perspective, in human behaviour.'

Mark Elgar, Professor, School of BioSciences, University of Melbourne

'Obligatory reading for anyone interested in gender equality at work or home – your views on sex differences will never be the same.'

Catherine Fox, author of *Seven Myths About Women at Work*

'I was amazed at how often Cordelia Fine seemed to read my mind. Time and again, whatever counterarguments I might muster were anticipated and refuted. That is the hallmark of a writer in tune with her readership. Fine lives up to her name – she is an extremely talented writer.'

Michael Jennions, Professor of Evolutionary Ecology, Australian National University

'There aren't many psychologists out there writing books that make me laugh out loud and want to stay up late reading, but Cordelia Fine does the trick ... Read this book because it's fun, but also because it's a great antidote to lazy thinking and entrenched sexism.'

Rebecca M. Jordan-Young, Chair of Women's, Gender, and Sexuality Studies, Barnard College, and author of *Brain Storm: The Flaws in the Science of Sex Difference*

'Full of witty gems you'll want to underline and read aloud, *Testosterone Rex* gleefully debunks myths about the nature of masculinity and femininity. Without denying science or ignoring evolution, Cordelia Fine shows how biology, far from limiting our possibilities, extends them.'

Marlene Zuk, Professor, Ecology, Evolution, and Behavior, and author of *Paleofantasy: What Evolution Really Tells Us about Sex, Diet, and How We Live*

'The next time someone solemnly explains to me why evolution has caused men to be competitive and women not, women to prefer childrearing and men to race cars and run corporations, men to be promiscuous and women coy, I plan to whip out my well-marked copy of *T. Rex* and cite the science that says they're wrong.'

Sharon Begley, author of *Train Your Mind, Change Your Brain*

'Exciting, eloquent and effective. Deftly weaving together research from anthropology, biology, neuroscience and psychology, Fine shows exactly why and how the myth of testosterone and maleness plays out, and why it is false. This book is not politically correct; it is good science.'

Agustín Fuentes, Professor & Chair, Anthropology, University of Notre Dame, and author of *Race, Monogamy, and Other Lies They Told You*

TESTOSTERONE REX

ALSO BY CORDELIA FINE

Delusions of Gender:
The Real Science Behind Sex Differences

A Mind of Its Own:
How Your Brain Distorts and Deceives

TESTOSTERONE REX

CORDELIA FINE

UNMAKING THE MYTHS OF OUR GENDERED MINDS

This edition published in the UK in 2018
by Icon Books Ltd, Omnibus Business Centre,
39–41 North Road, London N7 9DP
email: info@iconbooks.com
www.iconbooks.com

First published in the UK in 2017 by Icon Books Ltd

Published in the USA in 2017
by W. W. Norton & Company
500 Fifth Avenue,
New York, NY 10110

Sold in the UK, Europe and Asia
by Faber & Faber Ltd, Bloomsbury House,
74–77 Great Russell Street,
London WC1B 3DA or their agents

Distributed in the UK, Europe and Asia
by Grantham Book Services,
Trent Road, Grantham NG31 7XQ

Distributed in Australia and New Zealand
by Allen & Unwin Pty Ltd,
PO Box 8500, 83 Alexander Street,
Crows Nest, NSW 2065

Distributed in South Africa by
Jonathan Ball, Office B4, The District,
41 Sir Lowry Road, Woodstock 7925

Distributed in India by Penguin Books India,
7th Floor, Infinity Tower – C, DLF Cyber City,
Gurgaon 122002, Haryana

ISBN: 978-178578-318-0

Book design by Barbara M. Bachman

Printed and bound in Great Britain
by Clays Ltd, Elcograf S.p.A.

For Isaac and Olly

CONTENTS

But in addition to being angry, I am also hopeful, because I believe deeply in the ability of human beings to make and remake themselves for the better.

—CHIMAMANDA NGOZI ADICHIE,
"We Should All Be Feminists"[1]

INTRODUCING TESTOSTERONE REX

One memorable evening, I mentioned over the family dinner that it was time to get our newly acquired dog desexed. At this point I should explain that my older son has a strange, unchild-like interest in taxidermy. Thus, ever since this boisterous, loving canine entered the household, my son has been campaigning for the dog, after it dies, to live on not just in our hearts, but in a tasteful, formaldehyde-preserved pose in the living room. To my son, then, my remark about neutering offered the possibility of a stopgap until that day should come. Dropping his cutlery in excitement, he exclaimed, "We could have his testicles made into a key ring!"

A lively debate on the merits of this idea then ensued.

I share with you this intimate moment from Fine family life for two reasons. First, I wish to draw attention to the fact that—contrary to a prevailing view of the feminist as the kind of person who could think of no more inspiring and motivating a start to the workday than to unlock her office with a set of keys from which dangles a man-sized pair of testicles—I strongly vetoed my son's suggestion.

The second reason is that there is a useful metaphor here. A pair of testicles hanging on a key ring is bound to capture attention; to mesmerize. "That's some key ring you have there," people would politely comment. But what they would really mean is that in some important way your identity has been defined. Idiosyncrasies, complexities, contradictions, characteristics in common with those who *don't* have genitals on a key ring—all this fades into the background. *Who you are* is someone with a testicle key ring.

Biological sex can capture our attention in much the same way. We are spellbound by it; keep it constantly in the spotlight. This might seem perfectly appropriate. After all, sex categories—whether you have female or male genitals—are obviously fundamental for reproduction. Sex categories are also the primary way we divide the social world. When a baby's born, their sex is usually the first thing we want to know about them, and the last demographic information you're ever likely to forget about a person is whether they are male or female. Perhaps it's not surprising, then, that we often think of biological sex as a fundamental force in development that creates not just two kinds of reproductive system, but two kinds of people.[1]

At the core of this way of thinking is a familiar evolutionary story (aptly dubbed the "Biological Big Picture" by one of its sharpest critics, University of Exeter's philosopher of science, John Dupré.[2]) As we all know, the two parents of every human baby are owed grossly unequal debts for the miracle of life. According to my rough calculations, the mother is due more or less a lifetime of unwavering gratitude in return for the donation of a nice plump egg, forty weeks or so full bed and board *in utero*, many hours of labour, and several months of breast-feeding. But for the father, who by the time of birth may have supplied nothing more than a single sperm, a quick appreciative nod might well seem sufficient. This fundamental sex difference in biological investment in a baby means that, at least in some respects, in our ancestral past the sexes required different approaches to life to achieve reproductive success. This, of course, is the bottom line—indeed the only line—in evolutionary accounting. Men's much smaller minimum investment in a baby means that they can potentially reap huge reproductive benefits from having sex with many different women; preferably young, fertile ones. Not so for a woman. What most constrains her is access to resources, to help her care for her biologically expensive offspring.

And so, the various versions of this well-known account continue, a form of natural selection called sexual selection—arising from the edge that some individuals enjoy over others of the same sex when it comes to reproduction[3]—came to shape different natures in the sexes. Men evolved a promiscuous streak, and to be risk-taking and competitive, since these were the qualities that best enabled them to accrue the material and social resources attractive to women, and to turn that sexual interest into a reproductive return. A man could do *okay* by sticking with one woman, but those nice guys never hit the reproductive jackpot. For women, on the other hand, this kind of rapacious acquisitive behaviour would usually have had more costs than benefits. Some authors propose an evolved female strategy of opportunistic affairs with genetically superior men, during the fertile phase of the menstrual cycle, in a "good genes" grab.[4] But the ancestral women who most often passed on their genes were the ones psychologically inclined to play a safer game, more focused on tending to their precious offspring than diverting their energy towards chasing multiple lovers, riches, and glory.

All of this appears to be cool, dispassionate, unarguable evolutionary logic. Feminists can rail at the patriarchy and angrily shake their testicle key chains all they like; it's not going to change the fundamental facts of reproduction. Nor will it change the cascade of consequences this has for the brains and behaviour of modern-day humans. These effects, we're often told, apparently encompass activities well beyond our ancestors' wildest imaginings, like growing cell cultures in a science lab, or travelling at great speed in a metal tube on wheels. Consider, for example, how University of Glasgow psychologist Gijsbert Stoet explains the persistence of the gender gap in the science, technology, engineering, and mathematics (STEM) fields:

> People are often guided by their unconscious desires. In the stone age, it was useful for men to be hunters and women to

> look after babies, and nature has helped by encoding some of these skills in the hardware of our brain. That still influences how we think today.[5]

I have to say that none of the many mathematicians and scientists I know do their research in a way that brings to mind a caveman chasing a bush pig with a spear, but of course things may be done differently in Glasgow. And a similar link between the past and the inequalities of the present is made by the contributor to a Formula 1 magazine:

> A 21st century human has a stone-age brain.
>
> Stone-age humans of course did not participate in the FIA Formula 1 World Championship, but the rewards of survival and of course mating resulted in a male brain tuned for hunting, aggressiveness and risk taking.
>
> This has been shown in studies to be represented in the way males drive cars today. This is the reason why males have a higher number of fatal accidents on the road than females. Females were of course during the same period honed to raise and defend offspring. This of course all sounds deeply sexist but it is a combination of historical fact and recent scientific study.[6]

Quite so! How could it be sexist to merely report the objective conclusions of science? In fact, *are* there any sexists these days? Or are there just people who recognize that our brains and natures have been shaped by evolutionary pressures responsive only to reproductive success in our ancestral past, with no concern for the future consequences for the representation of women in Formula 1 world championships, or on corporate boards? After all, as University of California, Irvine, neurobiologist Larry Cahill observes:

> To insist that somehow—magically—evolution did not produce biologically based sex influences of all sizes and sorts in the human brain, or that these influences somehow—magically—produce little or no appreciable effect on the brain's function (behavior) is tantamount to denying that evolution applies to the human brain.[7]

Indeed, as the number of studies reporting sex differences in the brain pile up, the argument that sexual selection has created two kinds of human brain—male and female—seems to get stronger and stronger.[8] Could John Gray have been right after all when he claimed that men are from Mars and women are from Venus? Some scientists have argued that although average differences in the way males and females think, feel, and act may, on a trait-by-trait basis, be relatively modest, the accumulated effect is profound. "Psychologically, men and women are almost a different species," was the conclusion of one Manchester Business School academic.[9] Cahill, likewise, suggests that this compounding is akin to the way that many small differences between a Volvo and a Corvette—a little difference in the brakes here, a modest dissimilarity in the pistons there, and so on—add up to very different kinds of car.[10] Perhaps not coincidentally, one is a nice, safe family vehicle with plenty of room in the trunk for groceries; the other is designed to offer power and status.[11]

We certainly often behave and talk as if the sexes are categorically different: men like *this*, women like *that*. In toy stores, sex-segregated product aisles (real or virtual) assume that a child's biological sex is a good guide to what kinds of toys will interest them. Supposedly in keeping with sex-specific selection pressures in our evolutionary past, "boy toys" encourage physicality, competition, dominance, and construction. Meanwhile, the pink aisle, with its gentler offerings of dolls, domestic toys, and beauty sets, reinforces the twin pillars of traditional femininity: nurturance, and looking pretty.[12]

Some schools boast sex-segregated classrooms, grounded in the assumption that biological sex creates useful categories for pedagogical needs. For instance, the advertising tagline of a boys' school near me—"We know boys"—suggests that a state of deep confusion would arise were a girl to suddenly appear at the school one day, expecting to be taught. "But we know *boys!*" one imagines the teachers exclaiming in despair.

Many books likewise reinforce the message that *Men Are from Mars, Women Are from Venus,*[13] with other titles promising to explain why *Men Are Like Waffles—Women Are Like Spaghetti,*[14] *Why Men Want Sex and Women Need Love,*[15] *Why Men Don't Listen and Women Can't Read Maps,*[16] *Why Men Don't Iron,*[17] and even *Why Men Like Straight Lines and Women Like Polka Dots.*[18] (Straight lines *are* very unwelcoming, I find.)

And when it comes to the workplace, many "gender diversity" consultants take it for granted that biological sex provides a useful proxy for the skill sets employees bring to organizations. To increase female representation at senior levels, they recommend that employers "harness the unique qualities of men and women."[19] To have mostly men in senior management positions, this argument goes, is a bit like trying to sweep a floor with nine dustpans and one brush. Take a typical offering of this kind, *Work with Me: The 8 Blind Spots between Men and Women in Business,*[20] respectfully reviewed in *Forbes* and *The Economist.*[21] Here, authors Barbara Annis and John Gray argue that workers need to cultivate a "gender intelligence"—meaning a better understanding of men's and women's different perspectives and needs, and proper appreciation of the hard-wired female talent for communality, collaboration, intuition, and empathy that provide the perfect balance to men's intrinsically competitive, goal-oriented, and sometimes socially insensitive approach.

When we think of men and women in this complementary way, it's intuitive to look for a single, powerful cause that creates this divide between the sexes. And if you're thinking right now of a hormone beginning with the letter *T*, you're not alone. Testosterone has long featured prominently in explanations of the differences between the sexes, and continues to do so. For example, University of Cambridge neuroscientist Joe Herbert's recent book *Testosterone: Sex, Power, and the Will to Win* leaves readers at no risk of underestimating its potency:

> At the end of any discussion of the impact of testosterone on the history of mankind in all its wide-reaching and powerful complexity, a simple fact remains: without testosterone there would be no humans to have a history.[22]

Now *there's* a conclusion to inspire the reverence the testicle deserves . . . or at least until you realize that the same fact applies to oestrogen, carbon, and even that dullest of elements, nitrogen. But still—Sex! Power! The will to win! As Herbert explains, these are exactly the masculine qualities that, according to received wisdom regarding our evolutionary past, were so necessary for male reproductive success.[23] The testosterone surge in males during gestation is critical for the development of the male reproductive gear. The sustained increase in testosterone at pubescence brings about sperm production and secondary sexual characteristics like increased muscle mass, facial hair, and broad shoulders. Wouldn't it make sense if testosterone also made men *masculine*,[24] creating a psychological wedge that makes men like *this*, while its minimal presence in females helps to make women like *that*? As the hormonal essence of masculinity, testosterone would ensure that the desire for sex, the drive for power, and the will to win develop far more strongly in the sex for whom it was reproductively beneficial in our evolutionary past.

We all know what this means for sex equality in the workplace, given the much higher average levels of testosterone in men than women. Men's wider range of reproductive possibility means that "the entire life history strategy of males is a higher-risk, higher-stakes adventure than that of females," as one scholar put it.[25] So what does it mean for hopes of equality if testosterone fuels the appetite for adventure? *Of course*, we should value the special qualities that arise from women's low-risk, low-stakes approach to life. As world economies struggle to recover from the reckless risk taking that brought about the global financial crisis, commentators ask if there is "too much testosterone" on Wall Street,[26] calling for more senior women in finance. To a woman, after all, with the merest dribble of testosterone coursing through her bloodstream, subprime mortgages and complex credit derivatives will not have the same irresistible appeal. But here's the other side of that coin. If, thanks to the hand of evolution and implemented by testosterone, one sex is biologically more predisposed to want to take risks and get ahead, then it simply stands to reason that this is the sex that will be more eager to, say, take on the gamble of entrepreneurship, compete in Formula 1, or aspire to a powerful status that every day brings the heady possibility of barking the words "Jones—you're fired!" As Dupré explains the implications:

> If status-seeking is shown to be an adaptation for *male* reproductive success, we have finally located the biological reason for the much lower status achieved by women. Let's leave the men to pursue status while the women devote themselves to the important business of staying young.[27]

It's true that we don't, as a rule, tend to think that the scientific facts of nature dictate how things *should* be. Just because a scientist says that something is "natural"—like male aggression or rape—obviously doesn't mean we have to condone, support, or prescribe it. But that doesn't mean that science has nothing to contribute to societal debates or aspirations.[28] Although scientific claims don't tell us how our society *ought* to be, that being the job of our values, they can give us strong hints as to how to fulfil those values, and what kind of arrangements are feasible.[29] As Macquarie University philosopher Jeanette Kennett points out, if an egalitarian society isn't "a genuine possibility for creatures like us . . . then, on the basis that ought implies can, egalitarian prescriptions and ideals are undermined."[30]

If it's typically only in male nature to play with certain kinds of toys, to want to work in particular kinds of occupations, to be willing to make the family sacrifices, and to take the necessary risks to get to the top, then that surely tells us something about what kind of society it's reasonable to hope for and aspire to. Stoet, for example, takes pains to reassure that his conclusions about the lingering impact of our evolutionary past on girls' interest in biology or engineering "does of course not mean that women in modern society should stick with traditional roles." He emphasizes that people should be free to make counter-stereotypical career choices. But he is also of the view that this opportunity will never be taken up with any great regularity, and that initiatives to equalize women's participation in higher-paying STEM careers "deny human biology and nature."[31]

This statement reflects a heavy responsibility shouldered by those who take this view of the sexes: to be the messenger of unwelcome but necessary truths. The principle of sex equality—that no one should be denied an opportunity simply on the basis of the genitalia they happen to have stowed in their undies—is reasonably well entrenched in Western, contemporary societies. True, the members of gentlemen's clubs were apparently taking a long, deep nap when that particular shift in social attitudes and legislation took place; but most of us get it, and the principle is enshrined in equal-opportunity legislation. But if the sexes are essentially different, then equality of opportunity will never lead to equality of outcome. We're told that "if the various workplace and non-workplace gaps could be distilled down to a single word, that word would not be 'discrimination' but 'testosterone'";[32] that evolved sex differences in risk preferences are "one of the pre-eminent causes of gender difference in the labor market";[33] and that rather than worrying about the segregated pink and blue aisles of the toy store we should respect the "basic and profound differences"[34] in the kinds of toys boys and girls like to play with, and just "let boys be boys, let girls be girls."[35]

This is Testosterone Rex: that familiar, plausible, pervasive, and powerful story of sex and society. Weaving together interlinked claims about evolution, brains, hormones, and behaviour, it offers a neat and compelling account of our societies' persistent and seemingly intractable sex inequalities. Testosterone Rex can appear undefeatable. Whenever we discuss the worthy topic of sex inequalities and what to do about them, it is the giant elephant testicles in the room. *What about our evolved differences, the dissimilarities between the male brain and the female brain? What about all that male testosterone?*

But dig a little deeper and you will find that rejecting the Testosterone Rex view doesn't require denial of evolution, difference, or biology. Indeed, taking them into account is the basis of the rejection. As this book shows, Testosterone Rex gets it wrong, wrong, and wrong again. Contemporary scientific understanding of the dynamics of sexual selection, of sex effects on brain and behaviour, of testosterone-behaviour relations, and of the connection between our evolutionary past and our possible futures, all undermine the Testosterone Rex view.

There is no dispute that natural selection shaped our brains as well as our bodies. If there *are* any feminist creationists out there—it seems like an unlikely combination of worldviews—I can attest that I'm not one myself. But as the first part of this book, "Past," explains, the familiar "Biological Big Picture" version of sexual selection is now looking decidedly vintage. Decades of research in evolutionary biology have destabilized the key tenets once thought to apply universally across the animal kingdom, whereby arduous, low-investing males compete for coy, caring, high-investing females. The sexual natural order turns out to be surprisingly diverse, and we also bring our own uniquely human characteristics to the sexual selection story. For many years now, science has been rewriting and humanizing this evolutionary account: not much remains of the old tale at the heart of Testosterone Rex, as the first three chapters show.

"Past" razes old assumptions that universal principles of sexual selection inexorably gave rise to the evolution of two *kinds* of human nature, female and male. This clears the way for the second part, "Present," to continue to build the case for the same conclusion, beyond sexuality. Needless to say, these days we all agree that "nature" and "nurture" interact in our development. But in the interactionism of the Testosterone Rex perspective, biological sex is "a basic, pervasive, powerful, and direct cause of human outcomes."[36] Sex is fundamental, so that story goes. It is the timeless, unchanging seed from which either a male or female developmental programme unfurls. Experience plays a secondary role in the individual's developmental journey to a male brain and male nature, or to a female brain and female nature. Of course there is variability—not all men are identical, nor are all women. But amid all the "noise" of individual differences, a male or female "essence" can be extracted: characteristics of maleness and femaleness that are natural, immutable, discrete, historically and cross-culturally invariant, and grounded in deep-seated, biological factors.[37] Whenever we say that "boys will be boys," or accuse progressive interventions of trying to "go against nature" we invoke the assumption that there are such evolutionarily intended outcomes or "essences."[38]

But as Chapters 4 and 5 show, while the genetic and hormonal components of sex certainly influence brain development and function—we are not asexual blank slates—sex is just one of many interacting factors. We are an adapted species of course, but also unusually adaptable. Beyond the genitals, sex is surprisingly dynamic, and not just open to influence from gender constructions, but reliant on them. Nor does sex inscribe us with male brains and female brains, or with male natures and female natures. There are no essential male or female characteristics—not even when it comes to risk taking and competitiveness, the traits so often called on to explain why men are more likely to rise to the top.

So where does that leave testosterone? How does it create masculinity, if there's no one way of being a man, no common masculine core? Testosterone affects our brain, body, and behaviour. But it is neither the king nor the kingmaker—the potent, hormonal essence of competitive, risk-taking masculinity—it's often assumed to be, as Chapter 6 explains. So while it's probably fair to say that it really *was* mostly men who brought about the global financial crisis, the currently fashionable contention that "testosterone did it" and that therefore more "endocrine diversity" will save us,[39] is an excellent example of what happens when flawed Testosterone Rex thinking is applied to research and public debate, as Chapter 7 concludes.

So what should we make of—do with—this new and evolving scientific understanding of the relations between sex and society?

The final part of the book, "Future," looks ahead. The death of Testosterone Rex, and the arrival of its scientific successor, should transform how we think about the prospects for social change. No longer can we assume that to decree sex differences "biological," "innate," cross-culturally universal, or manifestations of sexually selected adaptations, is to pronounce us stuck with them, as the last chapter explains. So what, as a society, do we want?

NO DOUBT TESTOSTERONE REX will survive the savaging it receives in this book, and—like a taxidermied family dog that persists past its natural life span—continue to linger on in the public and scientific imagination. However, hopefully it will be left looking gravely injured. Or at the very least a bit nibbled.

But seriously, Testosterone Rex is extinct. It misrepresents our past, present, and future; it misdirects scientific research; and it reinforces an unequal status quo. It's time to say good-bye, and move on.

A NOTE ABOUT TERMINOLOGY

A WHILE AGO, MY YOUNGER CHILD GROUND TO A HALT IN A homework assignment because he wasn't sure whether to use the word "sex" or "gender" to describe a school-camp exercise in which every boy was paired with a girl.

"*Well!*" I exclaimed gleefully when he posed the question, quietly thrilling with excitement to have been presented with such a perfect teachable feminist moment. "That's a very interesting question, Olly. Let me try to explain." At these words, Olly's older brother let out a small gasp. If you can imagine the faces of onlookers had the Little Dutch Boy suddenly removed his finger from the hole in the dyke, you will have a rough idea of his expression.

Ignoring this look with quiet dignity, I began my sermon on the principles of terminology, but was almost immediately interrupted.

"Just tell me *which*, Mum," my son said impatiently. "I've got multiplication homework to do as well. Is it 'sex' or 'gender'?"

HIS UNCERTAINTY ISN'T SURPRISING. From the late 1970s, the word "gender" began to be used as a way of drawing a distinction between biological sex, and the masculine and feminine attributes and status that a society ascribes to being male or female. The idea was that by referring to "gender" you highlight the role of these social constructions—what society makes it mean to be male or female—in creating disparities between the sexes, as opposed to the relentless unfurling of biologically determined male and female natures.[1] But this approach was short-lived. From about the 1980s onwards, the word "gender" also began to be used in place of "sex" as a way of referring to whether an individual is biologically male or female, including even nonhuman animals.[2] These days, for example, surveys regularly ask you to identify your "gender," even though typically the expectation is that your answer will be based on whether you have a vagina or a penis, rather than any gendered psychic qualities or preferences. The person processing your credit card application will not appreciate it if, instead of checking one of the two boxes, you make annotations to the effect that in some ways your gender is male, but in other, no less important respects, it's female. This shift in usage has therefore robbed the word "gender" of its original meaning and function.[3] In its place, some feminist scientists now use terms like "sex/gender" or "gender/sex," to emphasize that when you compare the sexes you are always looking at the product of an inextricable mix of biological sex and social gender constructions.[4] But while this makes good sense (as Chapters 4 and 6 make clear), it's not particularly conducive to a smooth reading experience. For this reason, I use "sex" when referring to comparisons based on the categories of biological sex, and "gender" to refer to the social ascriptions.

In a second sacrifice of scholarly pedantry for the sake of readability, I use the word "promiscuous" (rather than more technical and precise terms like "polygynous," "extra-dyadic coupling," "polyandrous," and "multiply mating"), despite this being a term that is coming to be frowned upon in evolutionary biology.[5] While "promiscuous" is a highly value-laden term, no moral judgment whatsoever is implied by its application here. Not even for those slutty sandpipers featured in the chapter that follows.[6]

PART ONE

PAST

CHAPTER 1

FLIES OF FANCY

BACK IN THE MISTS OF TIME THAT THANKFULLY CAST A HAZE over my dating career, I became entangled with a man who drove a Maserati. When I let this slip to my mother, she responded in the unnaturally bright tone of voice she uses whenever, in deference to my technical state of adulthood, she wishes to disguise the fact that she thinks I have made a decision that will lead inexorably to disaster. "Fancy, a *Maserati*!" she exclaimed. "Does he have *many* girlfriends?"

The unsubtly implied connection has an interesting scientific history.[1] In the middle of the last century, the British biologist Angus Bateman carried out a series of experiments with fruit flies. They would eventually become the wellspring of a flood of claims about the psychological differences that have evolved between women and men. If you have ever come across the idea that men drive Maseratis for the same reason that peacocks grow elaborately ornamental tails, then you have been touched by the ripples of this landmark study. Bateman's research was inspired by Darwin's theory of sexual selection, which was a much debated subtheory within Darwin's widely accepted theory of natural selection. (Natural selection is the process whereby the frequency of different versions of a heritable trait change over time, due to some varieties of a trait leading to greater reproductive success than others.) Sexual selection theory was, in part, an attempt to make sense of the mystery of why the males of many species display extravagantly showy characteristics, like the peacock tail. These phenomena demanded an explanation because they were so awkward for Darwin's theory of natural selection. After all, if a primary goal of your life is to avoid being eaten by another animal, then a large, eye-catching, wind-dragging, feathered rear sail is not an asset.

Darwin's explanation drew on richly detailed observations of animals and their mating habits. (As one *Nature* journalist observed of that period of history, "despite the Victorians' reputation for prudishness . . . there were few places in the world where courting animals could escape the note-taking naturalist.")[2] These field studies gave rise to Darwin's famous observation in *The Descent of Man, and Selection in Relation to Sex* that the cause of males' deviation from the female form

> seems to lie in the males of almost all animals having stronger passions than the females. Hence it is the males that fight together and sedulously display their charms before the female.[3]

On the fighting side, more formally known as intrasexual competition, Darwin proposed that some characteristics (like an imposingly grand size or an intimidatingly large pair of antlers) are usually selected for more strongly in males. This is because these kinds of features increase a male's reproductive advantage by enhancing his ability to fight against other males for access to females. On the other hand, more whimsical characteristics—like a splendid plumage, a tasteful odour, or an intricate song—have their positive effect on reproductive success by boosting the male's appeal as a mate for the female. This dynamic is termed intersexual competition.

Darwin acknowledged that the pattern he'd described was sometimes reversed, with females being competitive or ornamented, and males appearing in the choosy, less spectacular style. But this was less common because, Darwin suggested, the challenge to be chosen usually fell more strongly on males than on females. He implied that this was due somehow to differences in the size and mobility of sperm versus eggs. But it was Bateman who, picking up on this idea and developing it, offered the first compelling explanation for why it is that males compete, and females then choose from among them.

The goal of his research was to test a prediction from sexual selection theory. Just like natural selection, sexual selection needs variation in reproductive success in order to work: if everyone is equally successful in producing offspring, there's no basis on which to weed out less successful individuals. If, as Darwin suggested, sexual selection acts more strongly on males, then this implies greater variation in the reproductive success of males than in females—that is, a wider range between the least, and the most, reproductively successful individuals. Bateman put this assumption to the test for the very first time.[4]

To do this, he ran six series of experiments in which male and female fruit flies (*Drosophila melanogaster*) were trapped together in glass containers for three to four days. At the end of this period, Bateman worked out as best he could how many offspring each male and female had produced, and from how many different mates. He needed considerable ingenuity to do this, since the discipline of molecular biology, that today brings paternity-testing kits to supermarket shelves, did not exist in the 1940s.

A screen-buff might be tempted to describe the solution he came up with as a cross between *Frankenstein* and *Big Brother*. Each fly in his series was inbred with a different, distinctive mutation: some with charmingly evocative names (like "Bristle," "Hairless," and "Hairy-wing"); others distinctly creepier (such as the miniature- or even no-eyed "microcephalous" fly). Each fly had one dominant mutant allele (one of the two copies of a gene) and a recessive normal one: meaning, as you might distantly recall from high school biology class, roughly a quarter of the offspring would end up with a mutation from both mother and father, a quarter from the father alone, and another quarter from the mother alone. (The last lucky 25 per cent of the offspring would have no mutations at all.) This principle of genetic inheritance enabled Bateman to estimate how many offspring each male and female had produced, and how many different mates a fly had enjoyed.

The outcome of Bateman's six series of matchmaking was the first scientific report of greater male variation in reproductive success. For example, 21 per cent of males failed to produce any offspring, compared with only 4 per cent of females. Males also showed greater variation in the estimated number of mates. But it was the linking of the two findings that became the basis of explanations for why males compete and females choose: Bateman concluded that although male reproductive success increased with promiscuity, female reproductive success did not. His critically important explanation was the now familiar insight that male success in producing offspring is largely limited by the number of females he can inseminate, whereas a female gains nothing from further pairings beyond a single one (since her first mate should furnish her with plenty more sperm than she needs).

Interestingly, Bateman's study was largely ignored for over twenty years.[5] Then his argument was expanded in a landmark paper by the evolutionary biologist Robert Trivers.[6] In this paper, the economics of egg and sperm production was made more explicit, being expressed in terms of the larger female investment of a big, costly egg compared with the male's minuscule contribution of a tiny, single sperm. Trivers also pointed out that the lopsided costs of reproduction can go well beyond sex differences in the size of the original contribution of gametes (that is, the egg versus the sperm) to include gestation, lactation, feeding, and protection. Any female readers who have themselves reproduced will, I'm sure, be inclined to agree with this point about the substantiveness of the female mammalian reproductive role. (My own personal understanding of this deep truth occurred during my first pregnancy, on reading an unhelpfully vivid description of childbirth as a physical feat comparable to a person making their way out of a car via the exhaust pipe.) The more highly investing sex—usually females—should therefore hold out for the best possible male, Trivers speculated, as the costs of a poor-quality mating are considerable. But males would do best to compete with other males in order to spread their cheap, mass-produced seed among as many females as possible. A follow-up implication, argued Trivers, is that males usually have less to lose and more to gain from abandoning existing offspring in pursuit of a new mate.

The Bateman paradigm, as it's sometimes known, was for a long time "the guiding principle and cornerstone for much of sexual selection theory." As University of Missouri–St. Louis evolutionary biologist Zuleyma Tang-Martínez puts it:

> Up until very recently, the unquestioned assumptions underlying the study of sexual selection have been that eggs are expensive while sperm are unlimited and cheap, that males should therefore be promiscuous while females should be very choosy and should mate with only the one best male, and that there should be greater reproductive variance among males (as compared to females) because it is males that compete for females and mate with more than one female. Since females are, presumably, mating with only one male, this means that some males may mate with many females, while others may mate with few or none. This reproductive variance is then responsible for the sexual selection of traits possessed by the more successful males.[7]

Indisputably elegant, Bateman's conclusions, elaborated by Trivers, enjoyed the status of universal principles for many years. They also became the bedrock of claims about evolved differences between women and men, in which peacock tails are replaced with Maseratis, corner offices, or big shiny trophies. Just replace the phrase "many females" with "many girlfriends" and "traits possessed by the more successful males" with "Maseratis possessed by the more successful males" and the dots are all connected. From this evolutionary perspective, a woman aspiring to high status is a bit like a fish aspiring to a bicycle.

But in the past few decades there has been so much conceptual and empirical upheaval over sexual selection in evolutionary biology that, according to one professor in that field I spoke to, the classic Bateman and Trivers papers are now largely cited for sentimental reasons. And startlingly, the first set of contradictory data we'll look at comes from Bateman's own study.

ALTHOUGH BATEMAN'S CONCLUSIONS tend to evoke images of the Playboy Mansion or well-stocked harems, it's necessary for the time being to return to Bateman's unsalubrious glass containers. It was only in our young century that, noticing that this (ahem) seminal paper had never been replicated, or apparently even subjected to close inspection, the contemporary evolutionary biologists Brian Snyder and Patricia Gowaty reexamined it. As they acknowledge, they returned to the study with many advantages that Bateman had lacked. These included modern computational aids, more sophisticated statistical methods and—perhaps I can dare to add?—fifty years of feminist insights into how cultural beliefs can affect the scientific process.[8] Like other modern critics of the Bateman study, Snyder and Gowaty expressed considerable admiration and respect for Bateman's "groundbreaking" study. But as they point out, given its "foundational nature," it was "important to know that Bateman's data are robust, his analyses are correct and his conclusions are justified."[9] As it happens, no such reassurance was forthcoming. Snyder and Gowaty's inspection revealed some significant problems.

For a start, as you'll recall Bateman used different mutant

Drosophila strains so that he could infer reproductive success from the particular pattern of abnormalities passed on to each offspring. If this method had you squeamishly pondering the grisliness-squared of a fly unfortunate enough to inherit both a maternal and a paternal mutation, you are one step away from a significant problem: these mutations turned out to affect offspring viability, and Bateman only counted surviving young in his tallies.[10] But if, on the other hand, a fly was more likely to survive because it had only one mutation, or none, then it could only be assigned, at best, to one parent. With the parentage of so many fertilizations completely or partially unaccounted for, this left considerable scope for error. While Bateman recognized this issue, Snyder and Gowaty quantified it. They noticed that in two-thirds of Bateman's series of experiments, his data indicated that males had produced more offspring than the females: a logical impossibility, since every offspring of course had both a father and a mother. In other words, the data had been biased towards counting the offspring of males.[11] This bias is important because the very point of the study was to compare male and female variance in reproductive success, yet the data were biased in ways likely to inflate estimates of the male variance.

Even setting aside this source of bias in Bateman's data, a vital problem remains, raised first by Tang-Martínez and Brandt Ryder.[12] While recognizing that Bateman's study was "ingenious and elegant,"[13] they also pointed out that his famous finding that only males benefit from promiscuity—immortalized into a universal principle—actually only applied to his last two series of experiments. For reasons that remain obscure,[14] Bateman analysed the data from the first four series separately from the last two, and presented them in two separate graphs. Remarkably, females *did* show greater reproductive success with a greater number of mates in the first four series, albeit less so than males. But in the discussion section of his article Bateman focused primarily on the results that fit the notion of competitive males and choosy females. As Tang-Martínez notes, this selective focus was then perpetuated by others:

> With a few exceptions, most subsequent researchers presented and relied only on the data from Bateman's series 5 and 6 (Bateman's second graph). General discussions of sexual selection, and even textbooks in animal behavior, almost always presented only the second graph and the discussion was limited to these results, usually as an explanation of why males are promiscuous and females coy and choosy. The results of series 1–4, and any discussion of increases in female [reproductive success] as a function of the number of males the female mated with, for all practical purposes disappeared from the literature.[15]

To see what the results looked like without the apparently arbitrary split among the experimental series imposed by Bateman, Snyder and Gowaty reanalysed data from all six series pooled together. As they drolly point out, if only Bateman had done so himself, he could have proudly laid claim to the first evidence of the reproductive benefits of female promiscuity! Reproductive success increased with number of mates for both females and males, and to a similar degree. Considered together with the bias towards counting fathers' offspring, they concluded "that there is no serious statistical basis in Bateman's data for his conclusion that the reproductive success of females does not increase with the numbers of mates females have."[16]

It probably goes without saying that it is something of a setback that Bateman's principles are contradicted by Bateman's data. Of course, evolutionary biologists interested in sexual selection weren't idly lolling around for decades on the grounds that good old Bateman had discovered everything they needed to know back in 1948. They were busy doing experiments, and contemporary research has identified many species to which Bateman's principles do appear to apply.[17] However, *Drosophila melanogaster* turn out to be just the beginning of a more complicated empirical story. By 2012, a lengthy table in an academic behavioural ecology journal listed thirty-nine species, from across the animal kingdom, in which research had established that female promiscuity brings about greater reproductive success.[18] And while in many of these species this link is nonetheless stronger for males, sometimes it's equal (for instance, in the yellow-pine chipmunk and the wild eastern salamander).[19]

This helps to explain why, contrary to the historical understanding that promiscuity is generally the preserve of males, it's now clear that female promiscuity is abundant across the animal kingdom—from fruit flies[20] to humpback whales[21]—and is "widespread" among primates.[22] This revelation owes a large debt to the DNA paternity-testing techniques that have enabled researchers to part the veils of discretion that previously obscured rampant female promiscuity (most particularly in many supposedly monogamous female birds).[23] Consider the lek: a mating arrangement in which males compete with each other in a specific territory or arena in a winner-takes-all conflict for sexual access to females. It is the paradigm case of competitive males and choosy females. But in some species, on closer inspection with the benefit of paternity-testing techniques, it has been turned upside down. For example, observations over two years of the buff-breasted sandpiper, a beautiful shore bird, suggested that in line with traditional expectations of how leks operate, one fortunate male was seen to be involved in 80 per cent of matings in the first year, and 100 per cent in the second.[24] Well worth *him* taking any risk to reach that top-bird position, you would think. But DNA paternity testing of over 160 offspring hatched during that time revealed that much had taken place out of sight. Far from one or two males having all the reproductive luck, at least fifty-nine different males had fertilized eggs in the forty-seven broods tested! (Eggs from the same brood can have different fathers.) This meant that "there were actually more fathers than mothers."[25] Recall, there's supposed to be only *one* father shared among the entire community of mothers. Moreover, most males only bore offspring with a single female, yet a remarkable 40 per cent of the broods had more than one father.

There could hardly be a greater contrast to traditional

understanding of how a lek operates. It's as if the women of a harem were to casually comment to the sultan when he popped by, "Oh no, that child's not one of yours—that's the second footman's daughter. . . . Eh? Ah, sorry. He's not yours either, that's the son of the chauffeur. Hang on, sultan, we'll find your kid. Nadia . . . *Nadia!* Do you remember which of these kids is the sultan's? Oh, yes, that's right. That boy over there playing with his half-brother. He's yours. Almost certainly."

In fact, there were already striking reports of female promiscuity even in the 1960s and 1970s, as University of California, Davis, behavioural ecologist Sarah Blaffer Hrdy has pointed out. Take the big cats, such as the lioness who might, during oestrous, mate as many as a hundred times a day with multiple lions. Or, consider savanna baboons, reported to actively seek numerous, brief pairings.[26] Yet somehow observations such as these failed to make much of a conceptual dent: perhaps because, as Hrdy wryly suggests, "theoretically the phenomenon should not have existed."[27] (As anthropologists like to quip: "I would not have seen it if I hadn't believed it.")[28]

Hrdy can lay claim to the best-known challenge to the notion of female monogamy. While studying a gray, black-faced species of langur monkey in India as a graduate student of Harvard University, she noticed that the females would routinely solicit mates other than "their so-called *harem-leaders*." As Hrdy describes her slow intellectual dawning:

> My own first glimpse of a langur, the species I was to spend nearly 10 years studying intermittently, was of a female near the Great Indian Desert in Rajasthan moving rapidly through a steep granite canyon, moving away from her natal group to approach and solicit males in an all-male band. At the time, I had no context for interpreting behavior that merely seemed strange and incomprehensible to my Harvard-trained eyes. Only in time, did I come to realize that such wandering and such seemingly "wanton" behavior were recurring events in the lives of langurs.[29]

Given the risks and costs of these "excess" matings (such as disease and predation risks for leaving the group, as well as the investment of time and energy that could be spent doing something else), this was behaviour that required explanation. (Hrdy suggested that it helped to leave open the father's identity, making it less likely that the mothers' offspring would be killed.) Since this justly famous scientific revelation, researchers have come up with all manner of ingenious suggestions as to the advantages female animals might gain from multiple mates. Since it only seems fair that women, too, should have access to evolutionarily flavoured "the-whisperings-of-my-genes-made-me-do-it" excuses for cheating on a partner, I provide a selection of these ideas here. Proposed gains of female promiscuity include genetic benefits, healthier offspring, and the opportunity to set up sperm competitions that weed out inferior specimens. It's even been suggested that females may have sex with several males in order to sabotage the reproductive success of rivals, by depleting local sperm stocks.[30]

If this last possibility sounds more ludicrously Dr. Evil than Mother Nature, this may be because of how effectively the Bateman principles obscured the notion of *female* competition.[31] For years, it was assumed that, since even the most mediocre female can achieve the very modest feat of getting herself fertilized by an eager male, every adult female will reproduce just about as well as the next. Females, then, would be under little selection pressure to develop traits that give them a reproductive edge over other females. But as Hrdy pointed out more than three decades ago, and ongoing research continues to confirm, a female's status and situation can have major repercussions for her reproductive success—particularly over longer time periods, during which discrepancies in male reproductive success may even out somewhat, as males take turns at being "king of the hill."[32] Among primates, for instance, low-ranking females' ovulation may be suppressed by nearby dominant females, or they may be so harassed by other females that they spontaneously abort. In the event that they do successfully give birth, their offspring are less likely to thrive and survive, thanks to inadequate food, harassment, or even infanticide at the hands of unrelated females. Gruesomely, these marauding females have even been known to eat the infants they kill. Meanwhile, in species like the chimpanzee, higher-ranking females reproduce at a faster rate, and their infants are more likely to survive, apparently due to access to richer foraging sites.[33]

Resources and rank matter for females. (Indeed, now might be a good moment to remind ourselves that the expression "pecking order" comes to us courtesy of hens.) Dominant female mammals have been found to get more and higher-quality food, better access to water or nest sites, and to enjoy reduced predation risks—thus, "improved reproductive success among dominant females appears to be widespread in a variety of mammal species."[34] Given everything it takes to gestate, lactate, and successfully see off one's young into the world—food, protection, maybe a nice little nest or privileged use of a feeding ground—this makes sense. Those better able to compete for material and social resources will be more likely to successfully pass on their genes to the next generation, and even—via the *quality* of those offspring, or inheritance of rank[35]—to the generation after that.[36]

In short, neither promiscuity nor competition are necessarily the preserve of male reproductive success.

And a third challenge to the intuitive force of Bateman's principles is that males can be choosy too. This, of course, makes no sense if you start from the assumption that, for them, mating comes at the rock-bottom price of a single sperm from a limitless supply. But this turns out to be a profoundly misleading way of thinking about the situation. Take, for instance, the presumed dizzying abundance and trivial cost of male sperm. As a number of scientists have pointed out, both observation and personal experience attest to the fact that males do not offer up a single sperm in exchange for an egg.[37] They instead produce millions of sperm at a time (in humans, on the order of two hundred million)[38] that luxuriate in the gland secretions that make up semen. While the situation varies from species to species, biologists have concluded that, in general, "the antiquated notion that males can produce virtually unlimited numbers of sperm at little cost is demonstrably incorrect."[39] Indeed, in one spider species, males run out of sperm after mating just once.[40] Nor may one ejaculation be enough to ensure fertilization, further running up the biological bill.[41] There are other costs to mating, too, beyond sperm. The males of many species provide "nuptial gifts," such as nutrient-rich sperm packages, captured prey, or even parts of their own body. And for any species in which coitus is any more elaborate than a brutally efficient collision of gametes, there will be costs of time and energy for courtship.

All in all, there are good reproductive reasons for the males of some species to be discriminating. Reviews on the topic provide the amateur animal behaviourist with many fascinating case studies that indirectly illustrate the principle that mating comes with a nontrivial biological price tag for males.[42] The males of some species (like the stinkbug and the bucktooth parrot fish) address the problem of sperm expenses in a Scrooge-like manner, grudgingly "tailor[ing] the size of their ejaculates"[43] to the reproductive quality of the receiving female.[44] Others, like the marsupial mouse *Antechinus*, take the opposite approach of splurging abandon, essentially mating to death during a brief breeding frenzy.[45] The price of sex for the male St. Andrew's Cross spider is so high that he only mates once. As the University of Melbourne evolutionary biologist Mark Elgar explained to me, this is because during this very special occasion "he foolishly breaks his copulatory apparatus and the female puts him out of his embarrassment by eating him."[46] (No wonder they're cross.) Other species keep costs down with self-imposed chastity. In Elgar's lab, male stick insects (Macleay's Spectre) are offered a mating opportunity every week. Despite apparently having nothing more demanding to do all day than resemble a stick, they only rouse themselves to take up this mating opportunity 30–40 per cent of the time.[47] Male mealworm beetles, Mormon crickets, and European starlings are similarly indifferent to female charms on a regular basis.[48] Indeed, it turns out that even male *Drosophila*, the original poster boys for the benefits of a philandering lifestyle, sometimes refuse the advances of willing females, presumably on the grounds that they're saving their sperm for the right partner.[49]

Given all the complications of the original Bateman story, it's unsurprising that there turns out to be no straightforward relation between parental investment and parental care either. For many years, people were so carried away by the dizzying reproductive possibilities of males that they forgot to ask where all the females-to-be-fertilized were to come from.[50] Overlooked was the fact that most of the females might already be busy with existing offspring. On average, male reproductive success can't outstrip that of females, due to the simple fact that every offspring has both a father and a mother. As evolutionary biologists Hanna Kokko and Michael Jennions point out, the theoretical possibility that a male could produce dozens of offspring if he mated with dozens of females is of little consequence if, in reality, there are few females available to fertilize, and competition for them is intense. As they put it, Trivers's parental investment theory

> implicitly assumes that the best response for males, who face more mating competitors than females, is to invest more heavily in weaponry, ornaments or other traits that increase their access to mates. There is, however, a valid counterargument: when the going gets tough, the smart do something else.[51]

A wonderful example is the horned dung beetle.[52] In this species, larger males grow long horns with which they belligerently guard entrances to the tunnels females use to mate and tend to their eggs. But while horned males wrestle at the tunnel entrances, smaller males take an easier approach that requires neither horns nor the exertions of battle. They simply sneak into a tunnel via a side entrance, find the female, and mate. (The females, incidentally, show no particular preference for their more traditionally masculine suitors.) In this case, males have one of two possible reproductive strategies, the smart "something else" approach being the one that sidesteps costly aggression and armoury. But in other species, males may evolve a more general pattern of doing "something else": paternal care. Whether or not paternal care evolves in a species seems to depend on the interaction of many different factors not yet fully understood. But certainly, it is much more common in birds and fish than in mammals, where gestation and lactation impose such huge biological start-up costs on the mother. Yet an exception to this are the primates, among some of which, at least, paternal care is common: "many males routinely protect, rescue, patrol, baby-sit, adopt, carry, shelter, feed, play with and groom infants."[53]

To be clear, the moral of all of this is not to try to argue that humans are really like buff-breasted sandpipers, stick insects, or chimpanzees. It's not to imply that senior female managers are suppressing the ovulation of their female interns, or to caution that, at some primal level, the women who work at the child care centre want to kill your toddler, and maybe eat him too. And the suggestion is certainly not that sex differences in reproductive roles are of no consequence. Rather, the point is the incredible diversity of sex roles across the animal kingdom: across species, biological sex is defined by gamete size but this, in turn, doesn't determine arrangements for mating or parental care.[54] This means that to question the popular Bateman-inspired view of human sexual relations isn't special pleading for humans to be exempted from foundational principles that apply to every other animal.

But no less important, even *within* a species, biological sex doesn't necessarily inscribe a fixed template for how the important business of reproduction should be achieved. Female bush crickets, for instance, are fiercely competitive when food resources are low, presumably because males supply them with nutrient-rich sperm packages. However, when the environment is abundant with the pollen they feast on, they switch to a more "conventional" choosy approach.[55] Who would have credited pollen with the power to flip sexual nature? Or consider the two-spotted goby fish, a species in which the ratio of available males to females changes rapidly over just a few months as males die off from the exertions of mating, parenting, and life in general. Again, this environmental change has a profound effect on mating. "Early in the season, males competed aggressively with each other for matings and were very active in courtship, whereas late in the season females . . . took over as the courting sex."[56] Then there is the dunnock (or hedge sparrow). In a book devoted to its habits, University of Cambridge zoologist Nick Davies observes that a mid-nineteenth century reverend and amateur ornithologist "encouraged his parishioners to emulate the humble life of the dunnock." Yet as Davies's fieldwork painstakingly documents, the hedge sparrow boasts "bizarre sexual behaviour and an extraordinarily variable mating system."[57] Depending on factors like female territory size, and how well both females and males are matched in fighting ability, dunnocks can wind up in a bewildering variety of sexual arrangements: monogamy, one female with two males, one male with two females, or two females sharing two males.[58] As Davies drolly notes, had the bird-loving Reverend's "congregation followed suit, there would have been chaos in the parish."[59]

In short, even within a species, biological sex doesn't necessarily determine mating strategies, which can instead "vary over time and space and are flexibly expressed as functions of ecological and social influences," as Swedish biologists Malin Ah-King and Ingrid Ahnesjö sum it up.[60] Parental care, they note, seems less flexible. But even this can sometimes vary within a species. For example, in some troops of wild Japanese macaque monkeys, adult males protect, carry, and groom one- and two-year old infants. But males from different troops, elsewhere in the country, show much less paternal care, or none at all.[61] Even when it comes to something as fundamental as mating, then, the effects of sex are more open-ended and flexible than we might tend to assume—a point we'll return to in the second part of the book.

So where does all of this leave us? In evolutionary biology, sexual selection is in an exciting state of turmoil; empirical revelations are turning accepted facts on their head, while conceptual changes are sending long-held assumptions flying out the window. A man with a Maserati is a fascinating phenomenon, deserving of study, to be sure. But whether he is the human biological equivalent of the well-antlered stag, his spotless luxury car the counterpart of the shimmering, biologically extravagant tail of the peacock—*that* is another matter altogether.

CHAPTER 2

ONE HUNDRED BABIES?

OF THE MANY BIRTH STORIES I HAVE HEARD, MY FAVOURITE is that of a woman—we'll call her Lily—from the mothers' group I belonged to. Lily's story begins in the usual way. She felt weary and nauseated in the first trimester, ate voraciously over the next three months while serious growth and consolidation took place *in utero*, then waddled around uncomfortably, becoming increasingly tired in the third trimester as the final touches to the baby were completed. Finally, Lily went into labour early. Not dangerously early, but inconveniently early, since her partner was overseas in the United States for work. Landing back in Melbourne in the nick of time after a sleepless twenty-hour journey worrying about Lily and the unborn child, he hurried to the hospital and was directed to the ward where her labours were finally drawing to their conclusion. He dashed to her side but, in a state of exhaustion and confronted there by the sight of a little pool of blood, he queasily lurched forwards onto the bed. Lily pushed him off her with some force. The father-to-be obligingly fell back, and cracked his head slightly on the unforgiving hospital floor. Lily's medical attendants flew immediately from her aid to his, and around the time she was pushing out their baby son, his father was tucked in a wheelchair, feeling the rush of cool air soothe his hot cheeks as he was hurried away to have his head tenderly ministered to.

My point, in case it's not already obvious, is this. When it comes to the miracle of bringing new life into the world, once a man has provided the ejaculate, even if his further contribution is to be merely useless, he is still doing better than some. This is why, at first glance, the reproductive potential for males appears to so easily surpass that of females. As psychologist Dorothy Einon points out: "In the time taken for a woman to complete the menstrual cycle that releases one ovum, a man could ejaculate . . . 100 times"[1] (although one hopes he wouldn't be so childish as to actually count). It's been estimated that, in what are described as "optimal" breeding conditions, a woman could bring about fifteen children into being in her lifetime.[2] Some individual women have even managed to give birth to many more than this: the anonymous first wife of a Russian peasant called Feodor Vassilyev had thirty-seven pregnancies yielding sixty-nine children. The highest recorded average rate, however, is ten to eleven children per woman, this being the impressive collective accomplishment of the women of the communal religious Hutterite group in the early twentieth century.[3] And, as is so often observed, a man could potentially produce ten times as many babies in a single year. This, we are often told, must inevitably make a difference to the evolutionary murmurings within. As Bradley University psychologist David Schmitt explains:

> Consider that one man can produce as many as 100 offspring by indiscriminately mating with 100 women in a given year, whereas a man who is monogamous will tend to have only one child with his partner during that same time period. In evolutionary currencies, this represents a strong selective pressure—and a potent adaptive problem—for men's mating strategies to favor at least some desire for sexual variety.[4]

The debt to Bateman in the chain of reasoning is obvious, in the implication that, for males, producing offspring can demand as little as a mere tablespoon of ejaculate and some modest, pleasurable exertion. But as we saw in the previous chapter, in many species the situation is decidedly more complex, with some of the long-held assumptions at the foundation of the Testosterone Rex account routinely overturned on closer inspection. So what about humans?

Consider, Einon posits, a woman who on average has sex once a week for thirty years. Now suppose she bears a generous brood of nine children. As you can easily calculate for yourself, on average she will have sex 173 times per child. And for each of the 172 coital acts that *didn't* lead to a baby, there was a partner involved, having non-reproductive sex. To explore what this means for any man trying to reach the benchmark set by Schmitt of scoring a century of infants in a year, it's worth following Einon in breaking things down to clearly see the schedule involved.

First, the man has to find a fertile woman. For the benefit of younger readers, it may be worth pointing out that throughout most of human evolution the Tinder app was not available to facilitate this. Nor, as observed in the previous chapter, was there likely to have been a limitless supply of fertile female vessels for men to access. In historical and traditional societies, perhaps as many as 80–90 per cent of women of reproductive age at any one time would be pregnant, or temporarily infertile because they were breast-feeding, Einon suggests. Of the remaining women, some of course would already be in a relationship, making sexual relations at the very least less probable and possibly more fraught with difficulties. Let's suppose, though, that our man manages to identify a suitable candidate from the limited supply. Next, he has to prevail in the intense competition created by all the other men who are also hoping for casual sex with a fertile woman, and successfully negotiate sex with her. Say that takes a day. In order to reach his target of one hundred women per annum, our man then has just two to three days to successfully repeat the exercise, ninety-nine more times, from an ever-decreasing pool of women. All this, mind you, while also maintaining the status and material resources he needs to remain competitive as a desirable sexual partner.

So what's the likely reproductive return on this exhausting investment? For healthy couples, the probability of a woman becoming pregnant from a single randomly timed act of intercourse is about 3 per cent, ranging (depending on the time of the month), from a low of 0 to a high of nearly 9 per cent.[5] On average, then, a year of competitive courtship would result in only about three of the one hundred women becoming pregnant.[6] (Although a man could increase his chances of conception by having sex with the same woman repeatedly, this would of course disrupt his very tight schedule.)[7] This estimate, by the way, assumes that the man, in contradiction with the principle of "indiscriminately mating," excludes women under twenty and over forty, who have a greater number of cycles in which no egg is released. It also doesn't take into account that some women will be chronically infertile (Einon estimates about 8 per cent), or that women who are mostly sexually abstinent have longer menstrual cycles and ovulate less frequently, making it less likely that a single coital act will result in pregnancy. We're also kindly overlooking sperm depletion, and discreetly turning a blind eye to the possibility that another man's sperm might reach the egg first. In these unrealistically ideal conditions, a man who sets himself the annual project of producing one hundred children from one hundred one-night stands has a chance of success of about 0.0000000000
00000000000000000000000000000000000000000000000000000000000000
00000000000000000000000000000000000000000000000000000000000000
000000000000000000000000000000363.[8]

One remedy for these low odds, you might think, is for men to restrict their sexual attention to ovulating women. Traditional wisdom holds that this is impossible since, unlike the females of other species, women don't advertise when they're in the business phase of their cycle. But with recent findings that, for instance, men find isolated characteristics of women (like the scent of their bodily secretions) more attractive during the fertile period of the menstrual cycle,[9] there have been suggestions that women's ovulation isn't so concealed after all. Whether this translates into behaviour, though, is questionable: a large-scale study of married women failed to find any evidence that sex was more likely during ovulation.[10] And while this does leave open the possibility of these subtle attractions having more of an influence on casual sex, as biological anthropologist Greg Laden points out:

> The fact that you have to do carefully controlled studies and then look very closely at the data to see a pattern like this (if it even exists) should not be ignored: If human males were primarily attracted to ovulating females and not very interested in non-ovulating females, then that would be easily seen and demonstrated.[11]

Regardless, timing one hundred seductions so precisely would normally be beyond demanding.[12] Even allowing that this remarkable feat could conceivably (sorry!) be pulled off, the chance of producing a hundred children is still only 0.0000000000000000000000000 00000000000000000000000000000000000000000000000000000000000000 000000000000000000000000000000748.[13] To put that number in a little context, a man's odds of being killed by a meteorite in his lifetime is 0.000004.[14]

And they say *feminists* are wishful thinkers.

It's not quite the case, then, that just outside the padlocked gates of faithfulness stretch endless richly fertile fields in which men can sow their seed. Among various hunter-gatherer societies, whose way of life is supposed to best reflect our ancestral past, the estimated maximum number of children a man can sire is twelve to sixteen: not so different from that of women (which is nine to twelve). This number *is* bigger in herder-gardener societies, increasing men's reproductive variance compared with that of women, and the variation is vastly larger in the intensive agricultural societies that enabled a few powerful and wealthy men to acquire massive harems.[15] But greater male reproductive variance seems unlikely to have been universal in our evolutionary history, being instead only seen in certain ecological, social, and economic conditions. It's not very easy to come by data providing good information about men and women's reproductive variance. However, a study led by University of St. Andrew's Gillian Brown compiled eighteen relevant data sets from across the globe and cultural spectrum, including both current and historical populations with a variety of mating systems. As one might expect, in polygynous societies (in which a small number of men have multiple wives), men had greater reproductive variance than women (sometimes substantially, in other cases more modestly). But importantly, this wasn't the case overall in the monogamous societies.[16]

In short, fathering anywhere remotely close to a hundred babies a year just isn't something that any old Stone Age Tom, Dick, or Harry could have achieved. (Indeed, a promiscuous man would need to have sex with more than 130 women just to have 90 per cent odds of outdoing the one baby a monogamous man might expect to father in a year.)[17] It would require the unusual alignment of conditions that enable a man to set up a well-stocked and expertly managed harem. Harems have "exceptional status"[18] in the nonhuman primate world, have of course only ever been available to a very small number of men in human history, and are unknown in hunter-gatherer groups that lack the necessary hierarchies of wealth and power.[19] (And, of course, treating women like property has become rather unfashionable in many parts of the world.) As University of Notre Dame anthropologist Augustín Fuentes warns:

> The use of unrealistic figures of potential male reproductive success is counterproductive because there is no evidence that in humans or other primates such a dramatic lifetime reproductive skew occurs with any regularity in any population studied. Using such assumptions as a jumping off point, even if hypothetical, lays an unrealistic baseline that can then be

> used to create a variety of scenarios, all of which are faulty given the erroneous basal assumption.[20]

Or to put it a little less academically: Best of luck, Evolutionary Psychology Fantasy Man.

Evolutionary Psychologists, by the way, certainly don't propose that men are only interested in no-strings sex, or that women only ever desire monogamy. One account from this intellectual stable, for instance, argues that both sexes deploy both short- and long-term "strategies," although to different degrees and geared towards somewhat different partner qualities.[21] But for much of our evolutionary history, sexual behaviour driven by "indiscriminate desires that lead to obtaining numerous sex partners in high-volume quantity," as Schmitt describes the "short-term mating strategy" ascribed to men,[22] would not have been a plausible or productive route to reproductive success. This should prepare us for what the evidence—as opposed to stereotypical caricatures—has to say about the sexuality of contemporary Western men and women. In *Challenging Casanova: Beyond the Stereotype of the Promiscuous Young Male*, Wake Forest University psychologist Andrew Smiler observes that "guys who sleep around meet our expectations; guys who are monogamous seem like exceptions."[23] Yet as Smiler goes on to explain, these beliefs are based on an inversion of reality.[24]

Needless to say, relying solely on what people report about their sexual desires and behaviours isn't ideal (although ethically preferable, obviously, to spying on them). Men and women tend to manipulate information (like pornography use and masturbation) differently in order to better conform to the sexual double standard.[25] In fact, a major headache for sex researchers is that men reliably report a larger mean number of other-sex sexual partners than do women. This is logically impossible, since heterosexual coitus requires the presence of both a woman and a man. This impossible discrepancy seems to be mostly due to men's inaccurate reporting, and their "greater tendency to report large, 'round' numbers of partners." Once people's tallies get to about fifteen partners, they tend to answer with "ballpark figures" ending in multiples of five (*Let's see, there was Suzy, Jenny, Malini, Ruth, . . . call it fifty*) and the discrepancy between the mean for men and women is larger in the oldest age groups, for whom memory is presumably most blurred.[26] Men's apparently inflated figures also inflate their variance: but needless to say, sexual selection can only act on the reproductive outcomes of *actual* sexual experiences, not fabricated ones.

Even when we take these self-reports at face value, the differences between the sexes are of degree, not kind. Certainly, on average men currently report a greater interest in casual sex than do women—at least within the not-very-broad-slice-of-humanity-across-time-and-place that has been surveyed.[27] But there isn't a sharp line dividing the sexes; nor is the Casanova model of male sexuality a good fit for the majority of men. Take the second British National Survey of Sexual Attitudes and Lifestyles (NATSAL),[28] based on a random sample of more than twelve thousand people ages 16–44.[29] Again, a grain of salt is required for these figures: 16- to 17-year-old men report 0.4 more total other-sex sexual partners on average than do women of the same age; 35- to 44-year-old men report 9 more, suggesting that those ballpark figures are becoming increasingly inflated over time. But despite this, the most common number of sexual partners for both women and men over the previous three months, the past year, and even the last five years, was just 1.[30] Over their lifetime, the median total number of partners was 6 for men, compared to women's 4. As these modest numbers suggest, only a small fraction of men reported having had 5 or more partners in the last year: about 5 per cent (compared with about 2 per cent for women).[31]

Of course, men might *want* to have sex with many different women, but not be able to realize their preferences. Yet even when men are asked how many sexual partners they'd *ideally* like, the answers are not vastly different from women's responses, and show a strong disinclination in men to take up the heroic to-do list required for a sufficiently high turnover of casual sex partners to have decent odds of theoretically outreproducing a monogamous male. The NATSAL survey found that the vast majority of both men and women ideally preferred to be in a sexually exclusive relationship: 80 per cent of men, and 89 per cent of women.[32] Within the eldest age bracket of the survey (a still sprightly 35–44 years of age), the gap was even narrower (86 per cent for men and 92 per cent for women). Touchingly, the vast majority of married and cohabiting men were perfectly happy with the idea of sexual exclusivity.[33] This rough similarity between the sexes in the theory of monogamy also seems to translate into practice, at least according to self-report. Large-scale representative national surveys find that husbands are only slightly more likely than are wives to report having extra-marital sex.[34] Nor should one feel especially pitying towards single women: while 78 per cent of the single women surveyed in NATSAL ideally wanted to be in a monogamous relationship, so too did 67 per cent of the single men.[35] Finally, contrary to what one might expect on the basis of the assumption that men supposedly strive for social status in order to gain reproductive opportunities, men in the highest social class were the most likely to prefer to be married with no other sex partners, and the least likely to want to exclusively devote their sexual energies to casual sex.[36]

There is, however, an infamous duo of studies that does seem to support the Testosterone Rex view of a stark contrast between the sexual natures of women and men. In these studies, conducted by Russell Clark and Elaine Hatfield, moderately attractive young male and female decoys were positioned around a college campus.[37] The decoys were instructed to approach people of the other sex and initiate a conversation by saying: "I have been noticing you around campus. I find you to be very attractive." This abrupt opener was followed with one of three propositions: "Would you go out with me tonight?" "Would you come over to my apartment tonight?" or "Would you go to bed with me tonight?" Men and women were equally likely to agree to a date (about 50 per cent). But although

69 per cent of men agreed to visit the woman's apartment and even more men agreed to go to bed with her, almost no women expressed interest in visiting a strange man's apartment, and precisely zero consented to sex. Similar studies in Denmark and France likewise found men to be far more likely to report interest in agreeing to an implicit or explicit invitation of casual sex.[38]

This study is often hailed as a "real" test of sex differences in promiscuous inclinations, as opposed to what people merely say about themselves. Perhaps so, and an actual sexual temptation in human form may well override what men merely think (or prefer to report) they don't want. However, it's worth pointing out that the experiment ended shortly after the unsuspecting—and presumably startled—participant made his or her reply. We don't know, for instance, how many women who agreed to go on a date might have ended up having sex.[39] Nor do we know how seriously men took these highly implausible sexual invitations, or whether those who accepted them would have followed through. So far as I can tell, there was no way of distinguishing between a "Yes, sure," meaning *Imagine, such is the power of my sexual magnetism that this entirely sincere woman of robust mental health wishes to take me, a complete stranger, to a secluded place to have sex* versus a "Yes, sure," meaning *Very funny, did your friends put you up to this?* or *This is weird, but I'll be polite.* In fact, in a later paper-and-pencil simulation of the same study (in which participants had the scenario described to them, and were asked to imagine how they would respond) that took away the awkwardness of the situation, men overall were disinclined to accept either sexual invitation.[40] Even in a slightly more plausible version of the scenario, in which the proposer claimed to be a fellow student and the offer was preceded by a brief, polite conversation, many men reported that they would be uninterested, on grounds such as "Too forward, kind of weird, [gave] me the sense that they have a screw loose," and "It takes more than one conversation to get in my pants."[41]

A second obvious objection is that what this study is *actually* primarily showing is women's lack of interest in being murdered, raped, robbed, or inflaming the interests of a potential stalker. (Indeed, the study authors, and others, make this point.)[42] In the paper-and-pencil simulations of the original studies, women often cited the creepy, dangerous, stalker-ish feel of the situation by way of reason for turning down the offer.[43]

All in all, then, while the "Would you go to bed with me tonight?" findings represent one of the largest sex differences ever observed in psychological research, and it demands explanation, chalking it up to fundamentally different female and male sexual natures may be premature. And recent work by University of Michigan psychologist Terri Conley and colleagues unravelling the factors driving this famous result illustrates a critical point: social realities mean that women and men in these studies are simply not participating in the same experiment. It's not just that the experiment as experienced by women entails inviting them to put themselves in a situation that, according to years of advice and warnings, is the very epitome of "asking for trouble."[44] Thanks to the sexual double standard, there are two further disincentives for women.

First, a woman accepting an offer of casual sex risks being seen both by herself and others as a "slut," as Clark and Hatfield point out. Some have dismissed the sexual double standard as a cultural relic in places like the United States. Certainly, attitudes can shift: sometimes remarkably quickly, as I discovered once when visiting the home of a university boyfriend. His father protested strongly against me sleeping in the same bedroom as his son, given our unmarried state. His wife listened respectfully, then suggested that if this was how he felt he had better get the ladder, climb up to the attic, find the camp bed, carry it down the ladder, clean it off, mend the wobbly leg, set it up in the study, find some bed linen and make it up for me. My boyfriend's father considered this for a moment and then concluded that, upon reflection, one *did* have to move with the times.

And times *have* changed, with some paper-and-pencil lab studies (usually with college students) failing to find evidence of the sexual double standard, or only within particular demographic pockets,[45] or for less conventional sexual activities.[46] But the double standard does emerge when researchers move beyond fictional vignettes and talk to people. An ethnographic study of college students, for example, "reported that the majority of students believed in heterosexual double standards and classified women into dichotomous categories of 'good' women or sluts."[47] As the ethnographer summarized the typical attitude of the male students:

> Men have the right to experiment sexually for a few years. There are a lot of female sluts out there with whom to so experiment. And once I have gotten this out of my system, I will then look for a good woman for a long-term relationship (or for a wife).[48]

"Slut" is, of course, a word for which there is no real male equivalent. As Concordia University's Emer O'Toole observes in her memoir *Girls Will Be Girls*, this provides a powerful implicit lesson in sexual moralities:

> I learned a plethora of words for women who had lots of sexual partners—slag, slapper, slut, floozy, tramp, tart, loose, easy, prozzy, bike, whore—and one for men: gigolo, which always seemed to carry an air of humorous accomplishment somehow.[49]

Likewise, the closest match reported in a study of students' linguistic cultures was "hoebuck,"[50] a slang term so benign that the first hit that came up in a Google search when I tried it was "Hoebuck Realty." When "Floozy Homes" becomes a viable name for a real estate business, we'll know the sexual double standard is *really* gone. Presumably, when assessing the potential reputational effects of casual sex, perceived cultural norms will weigh more heavily than one's own, apparently idiosyncratic, views.[51] And although relatively progressive university students don't *themselves* endorse the sexual double standard (although men reject it less enthusiastically than women), they do think that *others* do.[52]

Also easily overlooked is the risk to women from a different kind of sexual double standard: the very distinct possibility of the event not being all that one might hope for. A large-scale study of thousands of female North American college students found that they had only an 11 per cent chance of experiencing an orgasm from a first casual "hookup." While a policy of politeness requires the observation that orgasms aren't everything in a sexual encounter, women were six times more likely to enjoy hookup sex if they'd had one.[53] Follow-up interviews revealed why it was that women had such slim odds of reaching a climax. Students generally agreed that it was important for a man to be sexually satisfied in any context, and for women to be sexually satisfied in the context of a relationship. However, there was no perceived obligation to provide sexual satisfaction to a woman in hookup sex. While many men felt that bringing their girlfriend to orgasm reflected well on their masculinity, they often didn't feel the same way about hookup partners. One participant quoted by the study authors captured this sense of selfish entitlement particularly neatly:

> Another man told us, "I'm all about just making her orgasm," but when asked if he meant "the general her or like the specific her?" he replied, "Girlfriend her. In a hookup her, I don't give a shit."[54]

What if the strange man on campus inviting you to join him in bed that night was *that* guy?

From this, we can consider a couple of ideas. The first is that perhaps an updating of gendered norms of chivalry could usefully be made. Assumptions that men will open doors for women and pay for dates by default could be abandoned, and that solicitude and generosity be redirected to the bedroom instead. The second is that some of the gap between the sexes in enthusiasm for casual sex might close if the event left men sexually frustrated the majority of the time, but women almost invariably enjoyed full sexual relief.

Little surprise, in light of all of this, is that when Conley presented student participants with a hypothetical version of the Clark and Hatfield experiment, she found that they perceived the situation to be very different for propositioned women compared with propositioned men. Male proposers were perceived as more dangerous than the female ones,[55] and women predicted that they would be perceived more negatively overall, and as more promiscuous, socially inappropriate, and sexually desperate if they were to accept the offer than if they were to refuse.[56] For men, by contrast, accepting the offer was perceived to enhance, rather than damage, their reputation. The students also guessed that a male proposer was less likely to be a good lover than a female one, and less likely to provide a positive sexual experience[57]—apparently quite accurately, at least for North American student populations. These differences all made a difference to the likelihood of accepting the offer, with perceived sexual prowess of the potential partner being particularly key. Importantly, Conley found that this was the case not just for the frankly improbable Clark and Hatfield scenario, but also when it came to real offers of casual sex that participants had received in the past. And when the situation was modified to involve celebrities, or a close friend, rather than a complete stranger—a way of attempting to equalize the danger and pleasure perceived and anticipated by male and female participants—sex differences in interest in accepting the offer disappeared.[58]

Certainly, hypothetical paper-and-pencil tests of sexual behaviour are limited, and this isn't to present Conley's studies as the last word on the matter. Other research, for instance, found no evidence that men and women perceive different social risks from taking on multiple sexual partners, or that this contributes to sex differences in the desired number of sex partners.[59] Nor is the point that women's and men's sexuality is really just the same. But these studies perform a useful service in drawing attention to what appears to be easily overlooked: the many different social factors, still unequal for women and men, that feed into sexual decision making. Ironically, the need for this reminder was highlighted by a dismissal of Conley's findings by a prominent psychologist, on the grounds that females' interest in having sex with celebrities "may be motivated by more than sex."[60] As if sex, in the normal course of events, is separate from, and untouched by, identity, reputation, gendered norms, notions of "conquests" and "sluts," peer pressure and prestige, power, economics, relationships, culturally shaped sexual scripts, body shame, or any other complex part of one's inner and outer life.

This brings us to the important point (expanded on in the next chapter) that sexual behaviour viewed through the lens of the Bateman worldview filters out our humanity. Consider how Evolutionary Psychology–inspired researchers explain why attached men in their studies turn down offers of casual sex. Apparently obvious explanations—moral values, commitment, loyalty, simple lack of interest in having sex with someone who isn't the person they love—are ignored; instead, sexual restraint is explained in terms of reproductive outcomes weighed by "the risk of losing a primary partner with good reproductive prospects following the revelation of infidelity."[61] Sex stripped of everything human sounds more like . . . mating, and as we'll see in the next chapter, it's not clear how much of that humans actually do.

None of this, by the way, is intended as cheerleading for the notion that monogamy is really men's "natural" preference, or promiscuity women's.[62] As University of Minnesota evolutionary biologist Marlene Zuk argues in *Paleofantasy: What Evolution Really Tells Us about Sex, Diet, and How We Live*, evidence from a variety of sources suggests that humans have successfully paired and reproduced using all sorts of social arrangements, varying by time, place, and circumstance. "As with diet, as with exercise, as with all the other features of our biology that people want to make into a single 'natural' way—we don't have just one natural pattern of the sexes," she concludes.[63] Even polyandry (a woman with two or more husbands) is, in particular demographic and ecological conditions, seen more often than previously supposed among small-scale hunters and gatherers and foraging horticulturalists, across many parts of the world, suggesting that "polyandry may have existed throughout human evolutionary history." Interestingly, social groups are "apparently capable of instituting (and abandoning) fairly high rates of polyandry in a very short time frame."[64] In a piece aptly titled "Humans Are (Blank)–ogamous," University of Massachusetts Boston anthropologist Patrick Clarkin points out that although you'd think, "given the importance of sex and mating in evolution, that natural selection would have put a straight-jacket on it and given us a stricter blueprint to follow . . . that doesn't seem to be the case."[65]

SCIENCE HAS MOVED A LONG WAY from a Testosterone Rex view of sexual selection in which, in accordance with universal evolutionary design, sports cars are the peacock tails with which competitive men compete for fertile female vessels, laying the psychological foundations of sex inequality. As we saw in the previous chapter, decades of research in evolutionary biology has been reexamining and challenging the Bateman-inspired principles at the foundation of the Testosterone Rex view; from the supposed cheapness of sperm, to the assumed pointlessness of female competition. Gone are the days when commentators could pointedly refer to, say, the patriarchal dynamics of the elephant seal household, in discussions about humans. The old assumption that sexual selection has created near-universal sex roles—males mostly like *this*, females mostly like *that*—has been replaced with growing recognition of the diversity of courtship and parental roles both across and within species.

This across-species variability means that there is no universal template for how genetic and hormonal components of sex play out to affect brain and behaviour—a point we'll come back to in Chapter 4. And the within-species species variability in "sex roles"—think bush crickets, dung beetles, hedge sparrows, and most obviously ourselves—points to a no less important conclusion (that we'll return to later in the book). Sexual selection hasn't locked such roles into sex-linked genes and hormones, but allows for individuals to be profoundly influenced by their social, material, physical, (and in our own case) economic, cultural, and political circumstances. This is important because, as we saw in the Introduction, the implications of the Testosterone Rex view of the effects of sexual selection extend well beyond the bedroom. Ultimately, that old tale claims that it isn't just sexism and discrimination that sustains the glass ceiling—not completely. At the core of this inequality are the whisperings of evolution. To men, it murmurs *That's right . . . keep going, son. I know it may seem counter-intuitive to suggest that spending eighty hours a week in a science lab becoming increasingly pale and weedy, and possibly developing rickets, will make you more attractive to scores of young, beautiful, fertile women, but trust me on this.* Instead, to women, evolution is whispering *Are you sure all this hard work is worth it? Why not go home, invest more in the few kids you've got? Oh, and maybe brush your hair a little? It'll make it glossier—more youthful.*

But this old story is on its last legs, and it's time to give birth to its successor. As my mothers'-group friend Lily and her partner discovered, new arrivals don't always wait until everyone is perfectly ready to welcome them. So too, here. It doesn't matter whether you're cheering it on in the birthing suite, or rushing away in a wheelchair clutching your head. It's on its way.

CHAPTER 3

A NEW POSITION ON SEX

At one memorable point in proceedings, the Maserati-driving beau gave me a pair of Bulgari sunglasses. Perceived from the traditional sexual selection perspective, this was a brilliant strategic move: like the weaving of the intricate bowers with which male bower birds seduce female bower birds. It almost started to seem as though he had come into possession of a tattered old book titled *Making Sexual Selection Work for You: A Man's Manual*, and was following it closely. One Evolutionary Psychology perspective on consumer behaviour, for instance (with a head nod towards the superficially similar habit of male baboons in offering food to females in return for sexual access) suggested that "gift giving could have evolved as a distinctly male courtship strategy" that enables men to "flaunt their resources."[1] But although some writers apparently find the habit irresistible, within evolutionary biology it's generally considered rather bad form to attempt to explain the human condition by way of airy gesturing to superficially similar patterns in other animals.[2] Even among nonhuman animals, behaviours that look the same in two different species can have very different functions and evolutionary histories.[3] And while I don't pretend to be an expert in baboon psychology, I'm confident that a morsel of baboon fodder lacks the important weight of social meaning reflected in the expensive sparkle of Bulgari sunglasses. A recent analysis of gift giving in Nazi concentration camps, for instance, provides a compelling and moving illustration of how very un-baboon-like human gift giving can be. The main motivations to give gifts in this powerfully "identity-stripping context," the researchers concluded, were to assert agency, to form and reestablish social identities through relationships, and to restore a sense of humanity.[4] In humans, gifts "reveal an important secret: the idea which the recipient evokes in the imagination of the giver," as one scholar put it.[5] And how. British weather provides few valid opportunities for shaded eyewear, but even so, the Bulgari gift caused a collision of identities, held versus projected, of epic proportions. No member of my family had ever before owned a designer accessory and, for years, the sunglasses provided a rich source of amusement to my family. We all thought fondly of the man who so generously gave them. But we couldn't help but recognize the humour in what was, I'm sorry to say, a little like trying to attract a baboon with a peacock tail.

While nonhuman animals have their own trials and tribulations, this is just not something they have to worry about. The peahen doesn't wonder if the peacock's tail isn't perhaps a little *too* showy for her particular tastes and values; the male bower bird is free, I think, from anxieties that his bower doesn't reflect well on his prospects. Yes, we are animals, and we have evolved. But the uniquely human dimension we bring to everything we do, including the biological basics of birth, eating, excretion, and death, underscores how misleading it is "to assert the equivalence of, say, bird plumage and sports cars in attracting mates," argues University of North Carolina at Charlotte anthropologist Jonathan Marks.[6] The previous chapters disrupted the tight link in popular imagination between cheap sperm, vast reproductive potential, and an evolutionary drive towards a distinct male sexual nature. This chapter unmoors us altogether from the traditional view of sexual selection, with the idea that human sexuality is not only—perhaps not even primarily—about bringing together reproductive potentials. As Marks warns:

> To confuse human (cultural) sexuality and (natural) reproduction is classically pseudo-scientific. Of course sexuality is for reproduction—if you're a lemur. If you're a human, sexuality is far more than for reproduction; that is what evolution has done for human nature.[7]

Since he then suggests that "if you imagine sex to be biological, rather than bio-cultural, you're probably not going to have much of it"—read on.

IN A LONG AND THOUGHTFUL ESSAY, Macquarie University anthropologist Greg Downey argues that "in order to change popular understandings of evolution, we need not simply better data, but also better stories." His proposed alternative narrative to the "man-the-promiscuous-horny-hunter/woman-the-choosy-chaste-gatherer" story is a "long, slow sexual revolution," at the core of which is the understanding that "sexual expression in humans . . . has long been much broader than just to get gametes together successfully."[8] Importantly, this isn't a special pleading for humans to be considered outside of an evolutionary perspective. In fact, there's a compelling case to be made that sex doesn't serve purely reproductive purposes in other primates either.[9] The principle of "exaptation," whereby a trait that evolved for one function is redeployed for another, is now a standard one in evolutionary biology.[10] The textbook example even comes from the distinctly nonhuman characteristic of feathers, thought to have evolved first in dinosaurs for warmth, then for sexual display, and finally for flight in birds. Today, they continue to serve all three functions. John Dupré makes the point in his typically droll fashion, noting of his computer that "just because much of the underlying technology was developed with military applications in mind doesn't entail that my computer is constantly on the verge of planning a nuclear attack, or designing some instrument of mass destruction."[11] No doubt the initial function of our adaptive sexual desires and activity *was* reproduction, but this doesn't preclude it now having other functions. Sexual pleasure creates a "loophole in the evolutionary scheme," suggest Paul Abramson and Steven Pinkerton in *With Pleasure: Thoughts on the Nature of Human Sexuality*, which "permits sexual pleasure to be co-opted to other [nonprocreative] purposes, such as the facilitation of bonding and the reduction of personal and interpersonal tensions. . . . The intense pleasure that accompanies sex may serve to motivate copulation and thereby facilitate reproduction, but this is no longer its sole function."[12] This isn't the same as saying that humans sometimes have sex for reasons other than the conscious intent to reproduce, which is obviously true. One survey of students yielded no fewer than 237 distinguishable reasons for having sex,[13] my favourite of which is "I wanted to change the topic of conversation." (I've always wondered in which settings this is a motivation for sex. Dull dinner parties? Lab meetings that have turned to the awkward question of who forgot to order the pipettes?) Rather, the point is that its functional role goes beyond the merely reproductive.

Why this should have come to pass I will not attempt to explain, and for that I make no apology. Academic hypothesizing about human behavioural evolution reminds me of nothing so much as playing Pictionary with my dad. My father has many strengths, but none of them lie in the direction of the visual arts. Playing Pictionary, he doesn't so much draw a picture as happen to form a line or squiggle on the paper while maniacally gesturing with a pencil in his hand. (Although, technically, incorporating elements of charades into Pictionary is cheating, it is tacitly understood within the family that my father needs all the help he can get.) The researchers who speculate about the evolutionary origins of the human condition are, to my mind, in much the same position as someone teamed with my father in a Pictionary game, desperately trying to discern a meaningful picture from hopelessly inadequate information. *(It's fire! . . . No, you fool—surely that circle there is social complexity? . . . Or wait—could it be a big baby's head?)*

Fortunately though, there are several here-and-now clues to the non-reproductive purposes of sex in humans. Exhibit A we met in the previous chapter: the frequency of sexual activity even when there's no chance of reproduction. Given the costs and risks involved in sex, that doesn't make a lot of sense if the sole purpose is reproduction. In fact, for this very reason, in most animals hormones play a critical role in coordinating sexual activity, ensuring that sex only takes place when fertilization is possible. Why outlay the biological expense of souped-up secondary sexual characteristics and gamete production, or take the risks inherent in courtship, mating, and fighting, if there's no chance of reproductive success? If you are a male bird, for example, to sing an elaborate song of courtship that could attract the attention of a predator may only be a risk worth taking during the frenzied breeding season. In keeping with the same principle, outside of the breeding season when females are infertile and unreceptive, one might as well keep down biological costs by running a smaller size in gonads until spring is once again in the air. Human mating is conspicuously not like this. And even compared with other primates, in which sex is also released from hormonal control, our sexual activity stands out as especially unproductive.[14]

Exhibit B for non-reproductive sex is on a related theme: humans routinely engage in sexual pairings and acts, that not only often don't—but actually *can't*—lead to pregnancy. Women don't just have sex with men when they are not ovulating, but also when postpartum or postmenopausal. And sometimes, of course, it is not men with whom they are having sex, just as nontrivial proportions of men sometimes, often, or always prefer to have sex with other men. There are also many human sexual activities, like touching, kissing, and oral sex that likewise have no reproductive potential.

Exhibit C for a non-reproductive role for sex in humans is the absence of a penis bone in men, argues anthropologist Greg Laden, humans being the only ape for which this is so.[15] As a consequence, the efficiency of erection and orgasm is greatly reduced in men compared with most other apes:

> Male sexuality involves a much more elaborate, longer term, and complex set of psycho-sexual-social elements than usually found in apes, that are linked to social bonding. There are of course all sorts of exceptions, but typical, normal adult male human sexuality is actually somewhat complex and nuanced and not ape-like in many ways. Yes, folks, compared to *Pan troglodytes*, our nearest relative, human male sex is all about relationships.

Of course, we readily accept this when it comes to women's sexuality. In fact, Naomi Wolf brought the relational view of female sexuality to a whole new level in *Vagina: A New Biography*, claiming that

> his gazing at her, or praising her, or even folding a load of laundry, is not merely rightly thought of as highly effective foreplay; it is actually, from the female body's point of view, an essential part of good sex itself.[16]

Although I realize I have just observed that sex can take imaginatively non-reproductive forms, to include the folding of laundry does seem to take the thesis a little too far, for women and men alike. Certainly no one, to my knowledge, has argued for the critical importance of a tight bundle of carefully paired socks for successful male arousal, or the stimulating effects of the miracle that is the perfectly folded fitted sheet. So although it would have promised the easiest solution to date to women's unfair domestic burden, I suspect that it would be no small task to persuade men that although it may *seem* as though they are doing household chores, they are actually having sex. (*Honey, truly—this is the best sex I've ever had. Could you iron the tea towels too?*) But the considerable overlap between the sexes in their interest in a one-and-only sexual relationship (as well as in uncommitted sex) should dispel stereotypical contrasts in which only for women is sex about relationships. Indeed, in the previously mentioned survey of students' reasons for having sex, for both women and men, the top-rated reason was pleasure, followed closely by love and commitment.[17] As Andrew Smiler pleads:

> If we stop believing that boys and men are emotional cripples and fly-by-night Casanovas who just want sex, and start believing that they're full, complete human beings who have emotional and relational needs, imagine what might happen.[18]

Interestingly, even the apparent counterexample of the minority of men who purchase sex[19]—often taken as evidence of men's capacity and desire for purely physical sexual activity—turns out in some cases at least to be nothing of the sort. According to University of Leeds sociologist Teela Sanders, "a significant number" of men who purchase sex habitually or exclusively visit the same sex worker.[20] This seems surprising, given the natural assumption that the purchasing of sex is the manifestation of men's evolved desire for sexual variety, unencumbered by the restrictive relational obligations, moralities, and negotiations that sex usually entails. Why buy the same woman's sexual services twice, in a market exchange potentially as emotionally uncomplicated and uncommitted as getting one's car washed, or buying a bunch of bananas? Yet from her interviews with these men, Sanders concludes that

> commercial sexual relationships can mirror the traditional romance, courtship rituals, modes and meanings of communication, sexual familiarity, mutual satisfaction and emotional intimacies found in 'ordinary' relationships.[21]

Of course these "regular clients" are only a subset of men (and one certainly wonders how "ordinary" things seem from the perspective of the women providing these value-added sexual services). But Sanders's work indicates that even in this potentially most instrumental sexual exchange, for some men emotional intimacy, trust, communication, and familiarity are key parts of what is desired and paid for. Similarly surprising themes and motivations also emerged in an earlier small interview study of white, middle-class men who paid for sex, which found that an "attempt to structure the objective reality to be romantic/social continued for most of the individuals throughout the encounter." Interestingly, the researchers also reported that, in many cases, the transaction was followed, either immediately or in due course, by "a sense of disappointment and anticlimax." As one interviewee put it, in a striking reversal of stereotypical "morning after" roles:

> After the act one experiences a pang of feeling as if something is wrong because you just went through something which is not by any way, shape or form personal . . . there's absolutely no communication afterwards. It's over, finished. You are no longer of interest to the girls that you have just been with. And it's a big anticlimax afterwards.[22]

Or as one thirty-one-year-old man explained to Sanders:

> Sex is obviously quite an intimate act and it feels a bit funny just walking in with somebody you have never met before. Having sex with them and then walking out again. While seeing someone regularly it feels more like a proper human interaction.[23]

A proper human interaction. All this talk of "mating strategies"—the very term conjures up unfortunate images of people arguing around a boardroom table strewn with maps of local singles' bars studded with flags—obscures the point that we are "set up, psycho-sexually and physically, for non-reproductive sex," as Laden puts it.[24]

Once we stop viewing human sexuality through the narrow frame of simply bringing together two reproductive potentials,[25] it no longer seems so obvious and inevitable that men should strive for success while women fret about looking youthful. For example, Testosterone Rex reasoning holds that only a female's physical attractiveness is closely linked with her all-important fertility (indicated mostly by youthfulness, the physical correlates of which are taken to be more or less synonymous with female beauty). But from a purely reproductive perspective, there is good reason for women to also be drawn to good looks and dewy youth. Some Evolutionary Psychologists suggest that females have evolved a "short-term sexual strategy" in which they seek casual sexual encounters with men of good genetic stock, with attractively masculine facial and bodily features supposedly being walking advertisements for their superior genes.[26] What's more, as Hrdy pointed out some time ago, "older men . . . even if still potent, might deliver along with their sperm an added load of genetic mutations."[27] In line with this, recent research has established higher frequencies of "de novo" mutations (that is, those that arise for the first time in the gametes, rather than hereditary mutations) in the sperm of older men, and their contribution to genetic disease.[28] Presumably, then, the younger the man, the better the state of his "good genes." Yet despite all this, men don't wear uncomfortable platform shoes in order to make themselves look taller, rarely hand over fistfuls of cash to pay for major surgeries to make themselves more pleasingly *V*-shaped, or make their chins more handsomely prominent, nor line up in large numbers to have their foreheads paralysed with Botox injections. This absence of male enthusiasm for painful and expensive physical enhancements points to the possibility that deficiencies in reproductive potential can be, and are, forgivingly overlooked when it comes to sexual attraction.

Of course, physical attractiveness is a significant factor in sexual and romantic decision making, and it's not merely social convention that says we aren't looking our best in our eighties. But once released from the assumptions of the old sexual selection story, it becomes more reasonable to question whether men will always care more about physical attractiveness, while women focus more on resources. As one scholar points out, data regarding the first question "have been collected, by and large, from urban, middle-class, and often college-educated participants," who hail from "cultural and ecological environments that are evolutionarily novel: They are engaged in wage labor, involved in local, national, and global markets, exposed to mass media, and reside in relatively large populations."[29] Studies that have looked at mate preferences in small-scale societies with economies apparently more in keeping with those of our ancestral past—such as the Hadza hunter-gatherers of Tanzania[30] and the hunter-horticulturalist Shuar of Ecuador[31]—found little evidence that the sexes place different importance on the physical attractiveness of a partner. In the latter study, for instance, while a comparison sample from UCLA showed the "typical" sex differences in the importance of physical attractiveness, no such differences were seen in the Shuar participants.

And what about the Bulgari eyewear of supposed successful human couplings: male resources? As we saw in Chapter 1, it's a mistake to make the blanket assumption that a female's resources and status are irrelevant to her reproductive success. They can be of critical importance in mammals, including primates. As Hrdy argues (in a statement that, with a little tweaking, one could almost imagine taking place between mother and son in the drawing room of a Jane Austen novel):

> Clearly it makes evolutionary sense for males to select females not only on the basis of fecundity but also on the probability of producing offspring that survive. When intergenerational effects are likely to be important, males should also take into account female status, kin ties, or home range quality.[32]

It's certainly the case that cross-cultural studies reliably find that women care more about a potential partner's material resources.[33] But as Dupré points out:

> Given, first, that women in most societies have fewer resources and, second, that women often anticipate dependency on the financial resources of their mates, this is not an observation in obvious need of a deep biological explanation.[34]

Without doubt, early motherhood creates dependency on others. It's exhausting, time-consuming, and hungry work. But in an evocation of the female bush crickets from Chapter 1, that enjoyed the flexibility to adapt their mating strategy to their particular "economic" circumstances,[35] the greater the gender equity of a country, the smaller the gender gap in the importance of the financial resources of a partner (as well as in the importance of other preferences, like chastity and good looks).[36] Needless to say, the test case of a country in which the sexes enjoy economic equality doesn't yet exist. But even over the relatively short period between 1939 and 2008, preferences have shifted in step with a breaking down of traditional male breadwinner versus female homemaker roles, note psychologists Wendy Wood and Alice Eagly.[37] For men, the importance of good financial prospects, education, and intelligence in a partner has risen, while the importance of culinary and housekeeping abilities has decreased. Nor is men's self-reported interest in these "resource values" in women mere political correctness, being mirrored by changes in marital patterns in the United States. Whereas in the past, wealthier and better educated women were *less* likely to marry, now they are more so. As Wood and Eagly note, this means that women now enjoy "a marriageability pattern similar to that of men."[38]

In fact, we may be shortly waving farewell to the economics-inspired reproductive love story in which female Fertility Value meets male Resource Value, settles down, and maximizes reproductive success. In some cultures at least, to a far greater degree it seems that what we really want are partners similar to ourselves in these attributes. Behavioural ecologists Peter Buston and Stephen Emlen pitted the two perspectives—"potentials attract" versus "likes attract"—against each other. They asked close to a thousand U.S. college students to rate the importance in a long-term partner of the purportedly evolutionarily relevant categories of wealth and status, family commitment (presumed to be especially important to women), physical appearance, and sexual fidelity (supposedly particularly important to men in a partner).[39] The students then rated themselves on those same attributes. From a potentials-attract perspective, people with a high "mate value" (that is, highly physically attractive and sexually chaste women, and men of considerable wealth, status, and family commitment) will expect more complementary reproductive "potential" from a partner. But from a likes-attract perspective, people will want their partner to be similar to themselves: a woman who considers herself physically attractive and wealthy will desire something similar in a partner; a man who considers himself faithful and family focused will seek the same. Although if the researchers had only looked for data confirming the potentials-attract hypothesis, they would have found it and drawn the traditional conclusions—the likes-attract hypothesis actually won hands down in terms of its ability to explain people's preferences. For instance, a man's perception of his wealth and status was associated much more strongly with the importance he placed on the wealth and status of a potential partner than with her attractiveness. Similarly, a woman's self-perceived physical attractiveness had a much stronger effect on the importance she placed on a potential partner's looks than on his wealth and status.[40] Following a brief tour of data suggesting that more similar couples tend to have better-quality marriages, the researchers comment that their "results suggest that the emphasis should be shifted away from the standard approach that focuses on indicators of reproductive potential toward understanding how matching on a trait-by-trait basis contributes to marital stability and possibly to reproductive success."[41]

A later study, it must be said, failed to see evidence of likes attracting in a speed-dating situation, despite getting it on paper—a finding that highlights the somewhat dubious value of simply asking people what's important to them in a partner.[42] However, it's also possible that a speed-dating context may, by necessity, tend to push people towards focusing on an individual's most readily discernible qualities. Analyses of speed-dating data have found that, for men and women alike, physical attractiveness and youth dominate as predictors of a potential date's desirability.[43] But an analysis of actual matches made through an online dating Web site in China found that likes-attract again provided a much better explanation of the data than did potentials-attract. And even though there *were* signs here and there of "potentials" attracting too, sometimes this happened in the "wrong" way—for instance, there was evidence that, like men, "women also use their income to get more attractive men" and that "women with [a] better education background would like also to find a younger mate, just like men do."[44]

A relentless focus on "mating value," narrowly conceived, also contrasts with an analysis of several data sets reporting what characteristics men and women find more and less important in a partner. These show that for the past seventy-five years, across a number of different countries, the most important attributes in a long-term partner for both women and men have nothing to do with youthful fertility traded for resources. These most-desired attributes, in being unrelated to a person's reproductive worth, do not force commentators to propose what Dupré describes as "absurd evolutionary fantasies . . . in explanation of homosexuality."[45] These preferred characteristics do not offensively imply that the "mate value" of your wife—even if she happens to be the woman you love, the mother of your children, and the only person in the world who understands what you mean when you say someone had "'a beard like McFie's' or 'hair the same colour as that man in Hove who caught me kicking his cat'"[46]—is less when she's fifty than when she was twenty years younger. They are attributes that can't be bought, injected into you, or liposuctioned out of you. And they are also traits that have little to do with tax brackets, luxury European cars, or corner offices. Rather, they correspond to factors that reduce the chances you will want to throw a plate at your partner's head. They are dependability, emotional stability, a pleasing personality, and love.[47]

DOWNEY'S REFERENCE TO A "long, slow sexual revolution" tries to capture a fundamental feature of human sexuality. It wasn't suddenly, with the advent of the birth control pill in the last century, that human sexuality became unyoked from reproduction—that began a long time before. A broader understanding of human sexuality makes more visible the absurdity of "the tendency to argue that, in relation to sex, 'human nature' is what you get when you remove every human trait." To understand human sexuality, you can't simply "strip off everything that's distinctly human, like language, social complexity, and self-awareness,"[48] not to mention a person's politics, economic situation, social norms, and social identities. These are inextricably *part* of each person's sexuality.

University of Otago social historian Hera Cook provides a beautiful illustration of exactly this point in her rich account of the sexual revolution.[49] Cook notes that in eighteenth-century England, women were assumed to be sexually passionate. But drawing on economic and social changes, fertility-rate patterns, personal accounts, and sex surveys and manuals, Cook charts the path towards the sexual repression of the Victorian era. This was a time of reduced female economic power, thanks to a shift from production in the home to wage earning, and there was less community pressure on men to financially support children fathered out of wedlock. And so, in the absence of well-known, reliable birth control techniques, "women could not afford to enjoy sex. The risk made it too expensive a pleasure."[50] Victorian women turned to sexual restraint to control fertility, argues Cook, "a course of desperation that could be sustained only by imposition of a repressive sexual and emotional culture, initially by individuals of their own accord, and then . . . upon succeeding generations."[51] Cook describes the trajectory of Victorian women's sexuality from the mid- to late nineteenth century as one of "increasing anxiety and diminishing sexual pleasure."[52] Only with the increasing availability of reliable, accessible contraception in the early twentieth century was there a gradual relaxation of sexual attitudes and growing acknowledgement of the existence and importance of female sexual desire, culminating in the introduction of the birth control pill and the sexual revolution. For the first time in history, women were able to join men in sex without the risk of lifelong consequences.

Cook's rich perspective provides a useful reminder of the sheer newness, still, of the possibility of female reproductive and economic autonomy. So shouldn't we therefore see contemporary sexual relations as a particular point in a long, sexual revolution that is still taking place? Take, for example, the moral discomfort felt by Victorian couples that used the cervical cap as a contraceptive device. Since use of the cap suggested premeditated desire on the part of the woman, many couples considered insertion a "wanton act" and disapproved of it as an unfeminine "invitation to sexual intercourse," according to one birth control manual.[53] Even today, there are faded remnants of this attitude, in contemporary assumptions that female sexuality is passive and receptive, rather than the active author of its own desire: the coy female of the Testosterone Rex conception of sexual selection. But underscoring the point that a person's sexuality is exactly that—the sexuality of a *person*—a growing body of research (led by Rutgers University psychologist Diana Sanchez and her colleagues) suggests that an internalized notion of female sexual passivity can affect women's bodily sexual experience. For instance, heterosexual women with stronger mental links between sex and submission have greater difficulty getting aroused and achieving orgasm, and women who take a submissive role during sex experience less arousal (a correlation that isn't due simply to a lack of desire affecting both behaviour and sexual excitement). Their sexual dissatisfaction, in turn, reduces their partners' enjoyment.[54]

By contrast, women who endorse feminist beliefs report enhanced sexual well-being on several fronts—and not, apparently, simply thanks to the effects of those beliefs on men's propensity to fold the laundry. Feminist women are less likely to endorse old-fashioned sexual scripts, are more likely to have sex for pleasure rather than compliance, and enjoy greater sexual satisfaction thanks to a heightened awareness of their own desire.[55] What's more, women's feminism is good for the sexual satisfaction of their male partners too: a happy win-win situation.[56] In case you missed it, it was feminism that did that. I'm *just saying.*

By now it should be obvious that it's pointless trying to work out whether feminism reveals women's *real* sexually assertive nature, or is a social aberration that obscures their *natural* submissiveness. As Carol Tavris argues in her classic book *The Mismeasure of Woman*, "the idea that we have only to peel away the veneer of culture, the veneer of learning and habit, the veneer of fantasy, and the true sexual being will emerge" is profoundly mistaken.

> Our sexuality *is* body, culture, age, learning, habit, fantasies, worries, passions, and the relationships in which all these elements combine. That's why sexuality can change with age, partner, experience, emotions, and sense of perspective.[57]

This insight applies equally to the sexuality of men. It's worth pointing out that everyone, including Evolutionary Psychologists, recognize overlap in men's and women's sexual preferences and behaviour, and that these are responsive to social and environmental conditions. But once we stop trying to extract men's true sexual nature from the complicated social, economic, and cultural web in which every boy and man is embedded, the many hundreds of children begotten in Ismaïl the Bloodthirsty's vast and brutally guarded harem starts to look less like a manifestation of uncompromised, evolutionarily honed male sexual nature, and more a symptom of the fact that Mr. Bloodthirsty was a despotic asshole. In an article titled "The Ape That Thought It Was a Peacock," psychologists Steve Stewart-Williams and Andrew Thomas make the point nicely: "Does the behavior of these despots reveal the untrammeled desires of men in general, or does it just reveal the untrammeled desires of the kinds of men who become despots?"[58]

The ape that mistook itself for a peacock should also not forget that it's human.

PART TWO

PRESENT

CHAPTER 4

WHY *CAN'T* A WOMAN BE MORE LIKE A MAN?

The belief is all but universal that men and women as contrasting groups display characteristic sex differences in their behavior, and that these differences are so deep seated and pervasive as to lend distinctive character to the entire personality.

—LEWIS TERMAN AND CATHERINE MILES, *Sex and Personality*[1]

Men and women belong to different genders which are truly disparate.

—CLOTAIRE RAPAILLE AND ANDRÉS ROEMER, *Move Up*[2]

IN HER REVIEW OF WELL-KNOWN BIOLOGIST LEWIS WOLPERT'S recent book *Why Can't a Woman Be More Like a Man? The Evolution of Sex and Gender,*[3] psychiatrist and journalist Patricia Casey conveys profound relief at the challenge the book poses to the politically correct views of gender theorists. After briefly touring a litany of "naturally occurring differences hardwired into our genes," she concludes that the "obvious rejoinder" to Henry Higgins's famous question in *My Fair Lady*, "Why can't a woman be more like a man?" is "Because we aren't and we never will be."[4] This, apparently, is the only sensible conclusion to reach. After all, "the argument that testosterone and the Y chromosome have no influence on how we think and feel defies credibility."[5]

The assumption that, of necessity, these two biological agents of sex create not just a male reproductive system, but a distinctively male psyche, is in perfect keeping with the old view of sexual selection, whereby there's usually a strong and predictable link between being a prolific producer of cheap sperm, and a characteristically male way of comporting oneself. But as the previous section showed, even in nonhuman animals biological sex doesn't necessarily determine sexual nature, and especially not in ourselves. The biological realities of reproduction are never irrelevant, but even for dung beetles and hedge sparrows, other factors can have radical effects even on behaviour directly related to mating and reproductive success. These examples point to the surprising conclusion that biological sex may not be the fixed, polarizing force we often assume it to be.

In fact, even scientific understanding of sex determination (that is, how we come to be male or female) has shifted away from this view. According to the older, still prevalent, account, "The presence of a Y chromosome makes the embryo develop as a male; in its absence, the default development is along the female pathway," as Wolpert summarizes the process whereby sex is determined. The "key gene"[6] in this story is *SRY*, located on the Y chromosome. Individuals with the Y chromosome develop testes; in the absence of the Y chromosome, ovaries develop. The newly formed testes then produce high levels of androgens, particularly testosterone, which direct the development of male internal and external genitals; otherwise, female versions develop.

In this understanding of how maleness and femaleness come about, "The binary is stark: XX is female and XY is male,"[7] as Harvard University's Sarah Richardson observes. Reinforcing this cleanly binary view of sex is the apparently obvious state of the world. Even gender theorists, who unmoor themselves daily from reality as they grapple with the dizzying possibilities for reconstructing masculinity and femininity, agree that whatever genitalia you had when you put on your underwear in the morning, they will still be there when you get undressed again at night. About 98–99 per cent of the population *either* have XY chromosomes and male genitals (testes, a prostate, seminal vesicles, and a penis) *or* they have XX chromosomes and female genitals (ovaries, fallopian tubes, a vagina, labia, and a clitoris).[8] Tel Aviv University neuroscientist Daphna Joel refers to the three core markers of maleness and femaleness as genetic-gonadal-genitals sex: or 3G sex, for short.[9]

But this account, in which a person's sex hinges on the presence or absence of the almighty Y chromosome, turns out to be too simple. Consider, for instance, those few people in a hundred whose genes, gonads, and genitals *don't* all neatly align on either the male or female side: people you surely know, but quite possibly without knowing that you do. Social conventions, policies, and laws that require everyone to be either male or female obscure the biological reality that an "either/or" binary view of sex works for most people, but not everyone. A small but significant proportion of the population are "intersex": they are "like a female" in some aspects of 3G sex, but "like a male" in another, or are in between the male and female form in some aspect. For example, individuals with a male XY complement of chromosomes, but whose receptors don't respond to the androgens that are critical for masculinizing the genitalia, develop male testes but female external genitalia. As the Intersex Society of North America points out, this means that, despite the Y chromosome, these women "have had much less 'masculinization' than the average . . . woman [with XX chromosomes] because their cells do not respond to androgens."[10] Or consider congenital adrenal hyperplasia (CAH), in which an unusually large amount of androgens are produced *in utero*. In girls, this can result in somewhat masculinized external genitalia.[11]

Drawing attention to examples like these back in the 1990s, Brown University biologist Anne Fausto-Sterling risked making people's heads explode by observing that there are actually half a dozen or so sexes.[12] In the heyday of acclaim for the *SRY* gene on the Y chromosome as *the* sex-determining gene, Fausto-Sterling pointed out that intersex individuals are awkward for a model that doesn't allow for "the existence of intermediate states."[13] Following the neglected arguments of two geneticists, Eva Eicher and Linda Washburn, Fausto-Sterling suggested that deeply culturally embedded associations were implicitly at work in this scientific model: namely, "female," "passive," and "absence." The development of testes from an "androgynous" gonad is an active, gene-directed process, but in the absence of the potent, male *SRY* gene, ovarian tissue just . . . *happens*, as the default?

Contemporary sex determination science now recognizes that female development is as active and complex a process as male development. And what has also become clear is that *many* genes are involved in sex determination: *SRY* on the Y chromosome; a few on the X chromosome (including some involved in male sexual development); and then, surprisingly, dozens of others located on other chromosomes.[14] That's why, if you see the phrase "sex chromosomes" in scare quotes, it's not because some batty feminist scholar refuses to recognize the biological basis of sex, but because genetic sex *isn't* located in a stark binary—Y present or absent—but is scattered about the genome. Sex determination is therefore "a complex process." Rather than the simpler old scenario in which the *SRY* gene tips males onto a distinct developmental path, "the identity of the gonad emerges from a contest between two opposing networks of gene activity."[15]

Of course, when Patricia Casey rhetorically asks why a woman can't be more like a man, she doesn't wonder why it is that a woman can't turn her clitoris into a penis, or her ovaries into testes. Casey is expressing a common belief that sex—most prominently in the form of testosterone and the Y chromosome—has a fundamental effect on the brain and behaviour. As Penn State University psychologist Lynn Liben puts it:

> Males and females are assumed to have different "essences" that, although largely invisible, are reflected in many predispositions and behaviors. These essences are given—at the individual level—by a range of genetic and hormonal processes and—at the species level—by evolution. They are viewed as part of the natural order, likely to be presumed to operate across contexts and across the lifespan, and often presumed to be immutable (at least in the absence of herculean and unnatural efforts to change them).[16]

But does it make sense to expect sex to create essences in the brain and behaviour? Across species, the same evolutionary problem of sexual reproduction has been solved in lots of different ways—and this means that possession or absence of a Y chromosome[17] (and the other genetic components of sex) doesn't, in and of itself, dictate a particular way of behaving. But also, *within* some species—including our own, as this chapter fleshes out some more—neither sex has the monopoly on characteristics like competitiveness, promiscuity, choosiness, and parental care. The particular pattern, as we saw, depends on the animal's ecological, material, and social situation. This suggests that, even within a particular species, the effect of the genetic and hormonal facets of sex on brain and behaviour must not inflexibly inscribe or "hard-wire" particular behavioural profiles or predispositions into the brain; not even those more common in one sex than the other. Instead, they are drawn out to a greater or lesser degree, as circumstances dictate.

The sexes' reproductive roles (as in who produces which gametes, and puts what organ where) are distinct in a way that behavioural roles are not. Presumably this is because there is no environment or context in which having an intermediate version of the reproductive system, or putting together different parts of it in creative new ways—like a penis with a uterus, or testes with a set of fallopian tubes—would have been beneficial for reproductive success. Not so, though, for behaviour. None of which is to say that sex doesn't influence us above the collar. But should we expect the genetic and hormonal components of sex to have the same kind of effect on the brain and behaviour as they do on the reproductive system? With even *that* developmental process described by one expert as "a balance" rather than a binary system,[18] we might start to wonder not just whether, but *why*, sex should produce male and female brains, and male and female natures.

SEX CATEGORIES ARE THE PRIMARY way that we carve up the social world. It's the first thing we want to know when a newborn enters the world. It's often what we register first and fastest when we meet someone. We state it on almost every form we fill out. In most countries, we're legally required to be one or the other. We mark and emphasize it with pronouns, names, titles, fashion, and hairstyles.[19]

We probably wouldn't do this if maleness and femaleness—3G sex—didn't have certain important features. If last week you were female and had ovaries, a vagina, and so on, but this week you are male and have testes and a penis, those *M* and *F* check boxes probably wouldn't be nearly as common. If most of us were intersex in some way or another, the ubiquitous question, "Is it a boy or a girl?" wouldn't be so compelling. And if the shape of our external genitalia fell on a continuum, with the majority of people in an ambiguous midrange shape, it's an interesting question whether our sex would play such a key role in how we present ourselves to the world.

But of course 3G sex is not like this. The genetic and hormonal processes of sex, despite being complex and multifaceted, usually create distinct, consistent, and stable 3G sex categories. It's perhaps understandable for people to assume that sex has the same kind of fundamental effect on the brain as it does on the genitals. As Joel and a colleague put it, we assume that "sex similarly acts serially and uniformly, exerting an overriding and diverging effect, ultimately leading to the creation of two distinct systems, a 'male' brain and a 'female' brain."[20] It's not uncommon, in what passes for debate on Twitter, for people to counter the claim that there is no such thing as a "male brain" and a "female brain" by linking to a scientific article reporting a sex difference in the brain. In other words, as soon as we learn that brains differ according to sex, the implicit reasoning is that the brain must therefore also have a sex and, like the genitals, create female and male categories.

In fact, the classical scientific view proposed something along these lines. As with the genitals, testosterone was thought to be a key player, the prenatal gush produced by the newly formed testes masculinizing and defeminizing the brains of males in broad-brush fashion, while in its absence the brain is feminized. In this way, "genetic sex determines gonadal sex and gonadal hormones determine brain sex," as leading researchers Margaret McCarthy and Arthur Arnold neatly summarize it.[21] Scientists involved in non-human animal research assumed that these sex effects create discrete male and female neural circuits restricted to those involved in mating. But of course, for many psychologists and popular writers discussing the human condition, "mating behaviour" potentially includes in its scope just about every aspect of human psychology—from a visual system attuned to babies' faces, to a sense of humour that showcases one's superior reproductive potential.[22]

However, new evidence reveals a far more complicated picture, as McCarthy and Arnold explain. Sex isn't a biological dictator that sends gonadal hormones hurtling through the brain, uniformly masculinizing male brains, monotonously feminizing female brains. Sexual differentiation of the brain turns out to be an untidily interactive process, in which multiple factors—genetic, hormonal, environmental, and epigenetic (that is, stable changes in the "turning on and off" of genes)—all act and interact to affect how sex shapes the entire brain. And just to make things even more complicated, in different parts of the brain, these various factors interact and influence one another in different ways.[23]

For example, as Joel points out, environmental factors (like prenatal and postnatal stress, drug exposure, rearing conditions, or maternal deprivation) interact with sex in the brain in complicated and non-uniform ways.[24] Take just one study, showing that lab rats that have enjoyed a peaceful, stress-free life show a sex difference in the density of the "top-end" dendritic spines (these transmit electrical signals to the neuron cell body) in one tiny spot of the hippocampus. (The female dendritic spines are denser.) But look at the same brain region in a group of rats that have been stressed for just fifteen minutes, and now the dendritic spines of the male rats are bushy, like those of unstressed female rats. Conversely, the top-end dendritic spines of stressed female rats become *less* dense, like those of unstressed male rats. In other words, brief stress exposure reverses the "sex difference" for that particular brain characteristic.[25]

And it gets even more complicated than this. A particular environmental factor can have a profound effect on sex differences for one brain characteristic, but the opposite influence, or none, for others. For example, brief stress has a *different* effect on the "bottom-end" dendrites in this same brain region. Here, male and female dendritic spines are identical, so long as those rats have lived a stress-free life. But what happens if the rats are stressed? There's no effect on bottom-end dendritic spines in females, but their density increases in males. So what we have is a situation in which sparse top- and bottom-end dendritic spines are what you tend to see in non-stressed males and stressed females, bushy dendritic spines top and bottom is what you see in stressed males, and bushy tops with sparse bottoms is what you expect of non-stressed females.[26]

Confusing? That, in a way, is the point. You might also be beginning to wonder what exactly *is* the "male" pattern of dendritic spines? What is the "female" version? Unless you have very strong opinions on whether the true way of life for a laboratory rat is one of complete serenity, or whether it is the right and proper fate for every rat to experience brief episodes of high tension, there isn't really a good answer to this question. (For this reason, Joel recommends avoiding using the terms "male form" and "female form" to refer to brain characteristics.)

This particular study, conducted by neuroscientist Tracey Shors and colleagues, looked at one simple environmental effect on two extremely precise brain characteristics in one tiny part of the brain. Now imagine hundreds of these interactions between sex and environment, affecting many different features of the brain, as wild rats experience their own unique and rich tapestry of life. With each experience, some brain features change their form, others will not, giving rise to unique combinations of forms. What we should expect to emerge, then, from this "multiplicity of mechanisms"[27] is not a "male brain," or a "female brain," but a shifting "mosaic" of features, "some more common in females compared to males, some more common in males compared to females, and some common in both females and males," as Joel and colleagues conclude.[28]

This is exactly what Joel found for the very first time in humans, with colleagues from Tel Aviv University, the Max Planck Institute, and the University of Zurich.[29] They analysed images of more than 1,400 human brains, drawn from large data sets from four different sources. First, they identified around ten of the largest sex differences in each sample. Even this preliminary exercise challenged popular understanding in a couple of ways. First of all, contrary to the view that the brains of men and women are strikingly different, none of these differences were particularly substantial. Even for the very largest, the overlap between the sexes meant that about one in five women were more "male-like" than the average male. What's more, each data set had a different Top Ten list. As the authors point out, this shows that sex differences in the brain aren't simply due to sex, but depend on additional factors, the most obvious candidates being age, environment, and genetic variation.

Next, the researchers identified a "male-end" zone and a

"female-end" zone for each brain feature, based on the scores of the 33 per cent most extreme men and women, respectively. (An "intermediate" zone lies between the two.) Then they worked out whether people's brains are consistently on the male-end or the female-end of the continuum in each of these regions, or whether brains are a mix of male-end and female-end characteristics.

As the rat data might already have led you to expect, the results were decidedly in the favour of a mix. Between 23 and 53 per cent of individuals had brains with both male-end and female-end features (depending on the sample, type of brain measure, and method of data analysis). The percentage of people with only female-end or only male-end brain features was small, ranging from 0 to 8 per cent.[30]

So what *is* a "female brain" or a "male brain"?[31] Is a female brain the type of brain possessed by the very few individuals with consistently female-end brain characteristics—some of whom, by the way, are men? And if so, what kind of a brain do the majority of females have?

So sex does indeed matter, but in a complicated and unpredictable way. Although there *are* sex effects that create differences in the brain, sex *isn't* the basic, determining factor in brain development that it is for the reproductive system. Unlike the genitals, "human brains cannot be categorized into two distinct classes: male brain/female brain," Joel and colleagues conclude. Instead, they are "comprised of unique 'mosaics' of features."[32] One way to think of it is like this: a neuroscientist certainly might be able to correctly guess your sex from your brain, but she wouldn't be able to guess the structure of your brain from your sex.[33]

There is another important difference between sex differences in the genitals and sex differences in the brain. When it comes to the former, these very obviously serve males' and females' different fixed, timeless, and universal roles. Not even the most determined advocate of the view that a woman can do anything a man can would deny that a penis and testes function much better for delivering sperm than does a vagina and ovaries. But when it comes to the brain, "in many cases . . . the functions of neural sex differences are mysterious," point out neuroscientists Geert de Vries and Nancy Forger,[34] particularly the farther "up" the brain you get, away from the parts of the brain involved in very sex-specific functions, like ejaculation. Summarizing decades of endeavour in 2009, de Vries and a colleague note that

> hundreds of sex differences have been found in the central nervous system, but only a handful can be clearly linked to sex differences in behavior, the best examples are found in the spinal cord . . . we do not know the functional consequences of most of the others.[35]

This might come as a shock to some, especially given the willingness of some scientists and popular writers to conjecture links between sex differences in the human brain and complicated, multifaceted behaviours like mathematics, empathizing, or taking care of children.[36] But these speculations are, to put it politely, optimistic. There are no simple links between a specific brain characteristic and a particular way of behaving. Instead, how we think, feel, and act is always the product of complex assemblies of neural effort, in which many different factors act and interact.

To be very clear, the point is *not* that the brain is asexual, or that we shouldn't study sex effects in the brain. (Just for the record, I've never held that view.)[37] As several neuroscientists have argued, since genetic and hormonal differences between the sexes can influence brain development and function at every level (and throughout the brain, rather than just in a few reproduction-related circuits), investigating and understanding these processes may be especially critical for understanding why one sex can be more vulnerable than the other to certain pathologies of brain or mind. This, in turn, may offer helpful clues to potential causes and cures.[38] The point is rather that, potentially, even quite marked sex differences in the brain may have little consequence for behaviour.

This may seem counter-intuitive. In a keynote speech titled "When Does a Difference Make a Difference?" University of Toronto neuroscientist Gillian Einstein reflects on her puzzlement when she encountered this situation in her own research.[39] On the one hand, as she explains, there's clear evidence that oestrogens and progesterones can powerfully affect the growth, pruning, and connectivity of brains cells. Yet meticulous work in her lab identified little correspondence between oestrogen or progesterone levels (over the course of the menstrual cycle) on healthy women's moods—negative or positive. Contrary to popular myth and a thousand misogynist jokes—paging Mr. Donald Trump[40]—the important predictors of mood aren't time of the month, but stress, social support, and physical health.[41] Einstein had "a hard time with this," as she puts it. Intuitively, it stands to reason that "if you affect neurons, you affect the brain, and if you affect the brain, you affect mental states." Yet this wasn't what she found. Einstein's conclusion is that sex effects (like hormonal changes) have to be seen in the larger context of the many other neurochemical processes going on in the brain, and that "it takes a lot of neurons to mobilize a mood."[42]

This point about looking at the bigger picture brings us to a critically important possibility. What if the purpose of some sex differences in the brain were to *counteract other* differences? The numbers 3 and 2 are different from the numbers 4 and 1, but both combinations achieve the same additive result. Likewise, as University of Massachusetts Boston psychobiologist Celia Moore points out, different brains can reach the same ends via different neural means.[43] This is where evolutionary-flavoured preconceptions become key. For instance, in a *Psychology Today* blog subtitled "New Study Confirms That Men's Minds Come from Mars and Women's from Venus," University of Chicago psychobiologist Dario Maestripieri writes that, "from an evolutionary perspective, large differences in personality between the sexes make perfect sense."[44] From this conceptual starting place, what could make more sense than to suppose that any particular sex difference in the brain (or hormones) serves the purpose of making the sexes behave differently? It is easy to overlook the point that, to the extent that males and females need to be able to potentially behave in similar ways to get by in day-to-day life, evolution has to work out a way for this to be achieved in the somewhat different bodies with which they're bestowed. Humans, it's worth pointing out, rank pretty low on the Spectacular Bodily Sex Differences Scale. As Occidental College sociologist Lisa Wade points out, "If we were as sexually dimorphic as the elephant seal, the average human male would tower six feet above the average woman and weigh 550 pounds."[45] By contrast, beyond the genitals, women's and men's bodies overlap in everything from hormones to height, but there *are* average physiological differences. And so, bearing in mind that "male neural systems have evolved to control behavior most optimally in a male body and likewise for females," as de Vries and a colleague suggest, we can't assume that neurobiological sex differences always act to *create* differences in behaviour. Sometimes, they may in fact serve to iron them out.[46]

A beautiful example of de Vries's principle of "compensation, compensation, compensation"[47] comes from the neuroscience of birdsong, recently explained by Fausto-Sterling, in her book *Sex/Gender: Biology in a Social World*.[48] Songbirds represent one of the few success stories in linking sex differences in the brain to sex differences in behaviour. In the canary, for instance, the "song-control" brain region is bigger and denser in males, and this has been directly linked to males' superior singing. There's likewise a hefty sex difference in that brain region in another songbird species, the African forest weaver bird, being one and a half times larger in males. But unexpectedly, in *this* species, males and females sing together, in unison. How do both sexes sing the same song, despite this large sex difference in the song-control region of the brain? The answer: because of another sex difference. In females, the genes involved in song production areas "express" (produce brain-altering proteins) at a much higher rate than in males, compensating for their smaller neural real estate. "In effect," Fausto-Sterling explains, "the gene action advantage canceled out the size advantage leading to equal song production abilities."[49]

But still. These caveats and principles are all very well and good, but rats and other animals nonetheless don't have perfectly gender-egalitarian lifestyles. Yes, sexual differentiation of the brain is proving to be messier, more complex, and variable than previously appreciated. It's even less, well, sex-y than once thought, in the sense of sex being one of many interactive factors, rather than acting as a single, clear, predictable director of development. Yet somehow all of this still gives rise to certain kinds of behaviour that are more common in males, and other kinds of behaviour more common in females.[50]

This is a fair point. But easily passed over as a stabilizing buffer that allows the emergence of sex differences in behaviour is the environment.[51] Decades ago, Moore found that the high levels of testosterone in the urine of male newborn rats triggers a higher intensity of anogenital licking by their mothers, compared with the amount of licking received by female pups. This extra licking turns out, she found, to stimulate the development of sex differences in brain regions that underlie basic mating behaviour.[52] More recently, this more devoted maternal licking of males has also been linked to epigenetic effects in the brain and sex differences in youthful play behaviour, potentially a precursor to later sex roles.[53] In other words, the mothers' behaviour is an integral part of how male rats' brains and behaviour develop differently from females'.

This seems remarkable. *Maternal care*, a critical part of evolution's strategy for creating something as fundamental as male sexual behaviour? Shouldn't something so elemental be in the portfolio of the genes? But as developmental biologists have been pointing out for decades, offspring don't just inherit genes. They also inherit an entire "developmental system": an ecological legacy of place, physical environment, and structures; and a social legacy of parents, relatives, peers, and others who also provide important and reliable inputs as the animal grows and learns.[54] A rat will be born to a mother that will lick its anogenital region. A primate will be born in an environment with ready access to fruit. In other words, genetic material isn't the only source of developmental building blocks that can be relied on to be stable and enduring. So why not exploit this? Just as car engineers don't bother to design miniature crude oil distillers into cars since gas stations are readily available to motorists, "selection cannot favour a trait that compensates for the loss of a developmental input that is, as a matter of fact, reliably available," as University of Sydney philosopher of science Paul Griffiths explains. For example, primates have lost the capacity to synthesize vitamin C; why bother retaining this ability, since vitamin C is readily available on fruit trees?[55] Similarly, if a mother rat that enthusiastically licks one's anus and bottom is something that a male pup reliably inherits, along with his DNA, then natural selection will make use of it.

We humans obviously don't have anogenital licking as a means to provide a different developmental system for males and females. But the list of what we *do* have—a.k.a. gender socialization—is seemingly endless. No sooner has the determined gender scholar decided that she has completed her inventory, when Bic releases a special, slim pen "for her," or Oster creates an "Ironman" blender for males' very specific and distinct food-blending needs.[56] In newborns, small average sex differences in size, health, and ability to self-regulate might influence caregiver-baby interactions, even before parents' gender-related beliefs fully kick in, suggests Fausto-Sterling.[57] But it's the genitalia—and the gender socialization this kicks off—that provide the most obvious indirect developmental system route by which biological sex affects human brains. De Vries and Forger suggest that one way to think about these kinds of indirect pathways is that, ultimately, sex recruits a range of resources to reproductive-related ends. In other words, it delegates some of its developmental work to external contractors. And (as any home renovator knows), the bigger, longer, and more complicated a project is, the more scope there is for the end result to depart from the initial vision. And so, de Vries and Forger suggest, when it comes to humans, "with extensive social interactions and long development times, this means that there are plenty of opportunities to override or, alternatively, magnify the initial 'program.'"[58]

This is certainly one respectable view of gender socialization. Yes, we press dolls almost exclusively on girls; yes, we have a sexual double standard, and so on, and these social factors do make a difference. However, according to this view, these social norms exist because they reflect and respond to the original "programme" with which sex endows us: "nature" recruits "nurture."[59] But *is* there a particular "programme" or outcome that males and females are supposed to "develop to"?[60]

Some neuroscientists speculate that a benefit of environmental influences (like maternal behaviour) having a hand in sexual differentiation of the brain is that the process can therefore be tinkered with, in a way that's helpful for whatever the current environmental conditions happen to be.[61] And in our own species, this capacity is beyond useful: it's essential. The diversity of environments—and therefore conditions and roles—to which we need to have the potential to adapt far surpasses that of any other creature. Consider the variety of ways we feed ourselves, even: "It seems certain that the same basic genetic endowment produces arctic foraging, tropical horticulture, and desert pastoralism, a constellation that represents a greater range of subsistence behavior than the rest of the Primate Order combined," two evolutionary scientists note.[62] This critical difference between ourselves and other species is perhaps best illustrated by the reality TV show *Wife Swap*. In this long-running programme, viewers enjoy the mayhem that ensues when wives of generally very different social class, background, personality, and lifestyle swap homes, household rules, lives, husband, and children for two weeks "to discover just what it's like to live another woman's life."[63] I think I can say with low risk of the charge of anthropocentric bias that there is no other species in the animal kingdom for which this concept would work for seven seasons. Other animals are fascinating, to be sure. Many are highly flexible and adaptable. But there just aren't that many ways to *be* a female baboon. The unrivalled interest of human beings as objects of examination for reality TV programming reflects the fact that, as evolutionary biologist Mark Pagel puts it, we are "a single species with a global reach and ways of life as varied as collections of different biological species."[64] The anthropological, historical, psychological records and a single episode of *Downton Abbey* clearly show that how women and men behave "varies greatly depending on situations, cultures, and historical periods,"[65] as psychologists Wood and Eagly put it. We saw earlier in the book that even when it comes to something as basic as bringing the next generation into being, we humans have enjoyed an array of possibilities as to how to get the job done. A man might be a Chinese emperor with a large harem to service, or a contentedly monogamous British civil servant. A woman might be a mail-order bride, or actively seek multiple lovers in a socially sanctioned arrangement.[66] (Not even sexual preferences, despite being pretty important when it comes to the possibility of reproductive success, are reliably and exclusively other-sex oriented across people, times, and contexts.) It's simply not possible to designate any one way of life as representative of "male sexuality" or "female sexuality." So, too, for parental care: although greater maternal care seems to be universal across time and place, both mothers and fathers can be negligent and abusive, or loving and attentive, while cultural norms span from wet nurses to breast-feeding on demand, from boarding schools and thrashings to permissive, helicopter parenting. And as Wood and Eagly document, although it's universal for human societies to have a division of labour by sex, how those roles are shared, and what they involve, vary markedly across time, place, and circumstance, depending on the demands of the "cultural, socioeconomic, and ecological environment."[67] These very open-ended outcomes would be more easily achieved by a developmental pathway that runs from sex to socialization to the brain (and hormones, as we'll come to in a later chapter), rather than by an inflexible direct path from sex to brain.[68]

True, human societies' allocation of sex roles isn't always arbitrary: some roles are universally performed more commonly by one sex or the other. These, suggest Wood and Eagly, track physical differences between the sexes: in particular, men's greater upper body strength versus women's unique ability to grow and, until the invention of infant formula, feed babies. These make jobs like hand-to-hand combat and chopping wood more suited to men's physiques and, historically, jobs that required stretches of time uninterrupted by hungry babies would have tended to be allocated to men. But even these consistent sex role divisions aren't absolute. Sometimes, as Wood and Eagly describe, ecology and circumstances align in ways that bring about highly counter-stereotypical roles. For example, in some hunter-gatherer societies fathers show intensive care of infants, while in others women hunt large game, or hunt with dogs and nets,[69] or take part in military combat, including very occasionally in all-female units.[70]

One explanation might be that these represent cases of desperate times calling for desperate measures. Yet Wood and Eagly conclude that the "evidence that men and women sometimes engage in gender-atypical activities suggests a flexible psychology that is not vividly differentiated by sex."[71] This brings us to an evolving controversy: just how different—or similar—*are* women and men? A *Live Science* headline—"Men's and Women's Personalities: Worlds Apart or Not So Different?"—neatly captures the range of possible views. And of course the reason we argue about this more than, say, whether dogs are from Saturn and cats are from Neptune, is that it seems fundamental to the question of how we should feel about the status quo.

What is relatively uncontroversial (although the memo has yet to make it to a number of popular commentators) is that the majority of sex differences in the basic building blocks of behaviour—cognition, communication, social and personality traits, and psychological well-being—are relatively small. University of Wisconsin–Madison psychologist Janet Hyde drew attention to this important point in a now classic paper proposing the "gender similarities hypothesis."[72]

This was based on a synthesis of forty-six meta-analyses of sex differences in those fundamental building blocks. A meta-analysis is a statistical compilation of published and unpublished studies looking at a particular research question that, by pooling data, yields a more reliable estimate of what's going on. From this, researchers calculate a useful statistic known as the "effect size." It not only tells whether there is a difference between two groups but *how much* of a difference, if one exists. The bigger the effect size, the bigger the difference. A sex difference, after all, could mean anything from "almost all women scored higher than all men" (the situation when there is an effect size of about 3) to "there's a 56 per cent chance that a woman picked at random will score higher than a randomly selected man." This far less impressive difference, reflecting much greater overlap in the scores of women and men, would reflect an effect size of about 0.2.

What Hyde found is that more than three-quarters of the sex differences that emerged from these meta-analyses were either very small (0.1 or less) or small (0.35 or less), meaning that about 40 per cent of the time, *at least*, if you chose a woman and a man at random, the woman's score would be more "masculine" than the man's, or vice versa.[73] (If there were *no* average sex differences, this would happen 50 per cent of the time.) These included skills like mathematical problem solving, reading comprehension, and characteristics like negotiator competitiveness and interpersonal leadership style. A recent ten-year follow-up of Hyde's landmark paper that synthesised 106 meta-analyses of sex differences confirmed the gender similarities hypothesis no less emphatically.[74]

A more recent turn has been to look not just at the sex difference in a single variable but the pattern over sets of variables. In one recent study, Bobbi Carothers and University of Rochester psychologist Harry Reis found that people often score in a stereotype-consistent direction on some variables, but in the opposite direction on other, related ones. In other words, they can't be tidily sorted into "masculine" and "feminine" categories but are instead spread across a continuum.[75] As the researchers put it, "Although there are average differences between men and women, these differences do not support the idea that 'men are like this, women are like that.'" Instead:

> These sex differences are better understood as individual differences that vary in magnitude from one attribute to another rather than as a suite of common differences that follow from a person's sex.[76]

Another evolving aspect of this debate is that old arguments that sex inequalities are explained by women's intellectual inferiority have shifted towards claims instead that these inequalities are due to sex differences in values and interests. It's not that a woman *can't* behave like a man; it's just not in her nature to want to. Yet contrary to the Testosterone Rex perspective, sex differences in both "masculine" values (social status, prestige, control and dominance of people and resources, and personal success) and the "feminine" value of caring for loved ones are also small.[77] Nor are such priorities set in stone. The Pew Research Center in the United States, for instance, recently reported that young women have now overtaken men in the importance they place on success in a high-paying career, and the sexes are equally likely to count being a good parent and having a successful marriage as more important than lucrative workplace success.[78]

But perhaps, you may be thinking, women's equal work ambition is all very well and good, but only men tend to have the requisite ruthlessness to get ahead. A well-worn story is that, since evolutionary dynamics dictate that the nasty guy dominates and therefore gets the girl, men tend to be inherently more aggressive than women. In fact, there are some serious question marks hanging over this chain of assumptions.[79] But even setting these aside, the argument doesn't really work. The largest sex difference in aggression is, unsurprisingly, in the physical variety. (You'd have about a two in three chance of correctly guessing whether someone was male or female, based on whether they were below or above average in physical aggressiveness.)[80] But here are two facts about your own occupation that, with a few exceptions, are almost bound to be true. First, men predominate at the most senior or prestigious levels. Second, they didn't get there by virtue of their greater propensity to punch people in the nose. That's not to say that everyone plays nice at work. But meta-analytic findings for sex differences in verbal aggression fall into the very small to moderate range. (And in some places, like the Gapun village in Papua New Guinea, the typical sex difference is reversed, with women renowned for their colourfully aggressive tirades towards those who have displeased them.)[81] As for indirect aggression—the aim of which is "to socially exclude, or harm the social status of, a victim"[82] without getting blood on one's suit—if anything, the scale is tipped towards greater female aggression.[83] In short, sex differences in aggression don't get us very far in explaining the occupational status quo.

True, sex differences in occupation-related interests are larger. (I took a close look at the evidence supposedly showing the "hard-wired" basis of this in my previous book, *Delusions of Gender*.) According to compiled findings from one much-used inventory, more than 80 per cent of men report greater interest in "things" than the average woman, who has a greater fascination with "people"-inclined activities,[84] and this seems to be reflected in the kinds of occupations into which women have made the least inroads over the past three decades.[85] However, it's worth noting psychologist Virginia Valian's observation that simply labelling a dimension "things" or "people" doesn't make it so. For example, the three subscales of the inventory that make up the "thing" dimension require "thing" to be interpreted so broadly—including "the global economy, string theory, mental representations, or tennis"—that the term becomes "vacuous."[86] Valian also suggests that preconceptions about which sex *does* stuff with *things* have influenced the creation of the items. Why, for instance, don't activities like "Take apart and try to reassemble a dress" or "Try to recreate a dish tasted in a restaurant" appear on such scales?[87] But also, as Valian observes, the sexes are artificially divided when they are categorized as *either* "thing people" *or* "people people." In fact, being interested in things doesn't stop you from being interested in people, and vice versa. Many men and women are, of course, interested in both and would be pretty awful at their jobs if they weren't. For instance, I wouldn't care to have blood taken by a nurse, however sympathetic, who was completely uninterested in the mechanics of the syringe. Nor would I want to hand over the renovation of my house to a builder who had no interest in understanding or managing the delicate psychology of the tradesperson.[88]

One counter-response to claims that a person's sex tends not to be a very good guide to whether they will be "masculine" or "feminine" on a particular trait is that these usually modest differences nonetheless add up to something rather substantial. The neurobiologist Larry Cahill, as we saw in the Introduction, suggests that the argument that the sexes are similar because most differences are small is "rather like concluding, upon careful examination of the glass, tires, pistons, brakes, and so forth, that there are few meaningful differences between a Volvo and a Corvette."[89]

However, there's a problem with this line of reasoning. For many decades, researchers supposed that masculinity and femininity are polar ends of a single dimension: someone high in masculinity is therefore necessarily low in femininity, and vice versa. In fact, this assumption was built into the very design of the first systematic attempt to measure masculinity and femininity: a brisk 456-item questionnaire with the carefully obscure title, The Attitude Interest Analysis Survey.[90] The survey yielded a single score that placed each individual at a particular point on a single masculinity-femininity line. So if, for example, you felt that the word "tender" went most naturally with the word "loving" or "kind" then you lost a point (naturally!) for being feminine. By contrast, if your mind leapt unsentimentally from "tender" to "meat," then you may have had trouble getting second dates, but you did at least gain a point for being masculine.

It wasn't until the 1970s that this assumption was overturned by the development of two new scales.[91] Still in use today, these separately assess "masculine" traits of "instrumentality" (qualities like self-confidence, independence, and competitiveness) and "feminine" traits of "expressiveness" (such as being emotional, gentle, and warm and caring towards others). This revealed that it is possible to be both instrumental *and* expressive, or neither. To put it in terms of Cahill's car metaphor, one can have the safety, reliability, and room-in-the-trunk-for-the-weekly-groceries-ness of the Volvo *and* the power, status, and thrill of the Corvette. Or—and here I stress to any Volvo and Corvette owners that the comments that follow are made solely for pedagogical purposes—one can have the sluggishness of a Volvo *and* the expense of a Corvette. But even this two-dimensional model of gender is now known to be too simple. Correlations *among* masculine traits and *among* feminine ones are often weak or nonexistent. Having one masculine trait doesn't imply you have another, and likewise for feminine traits.[92]

In other words, differences between males and females may not "add up" in a consistent way to create two kinds of human nature; but rather, as with sex differences in the brain, create "mosaics" of personality traits, attitudes, interests, and behaviours, some more common in males than in females, others more common in females than in males. Joel and colleagues tested this idea, drawing on three large data sets, and using the same approach as they did for brains. Even looking at only twenty-five behaviours with at least moderate sex differences (this included attributes like communication with mother, being worried about one's weight, and delinquency, as well as strongly sex-stereotyped activities like playing golf and using cosmetics), between 55 and 70 per cent of people (depending on the sample) had a mosaic of gender characteristics, compared with less than 1 per cent who had only "masculine" or only "feminine" characteristics.

This makes the notion of female natures and male natures as unintelligible as that of female brains and male brains. Which of the many combinations of characteristics that males display should be considered male nature? Is it a profile of pure masculinity that appears to barely exist in reality? What does it mean to say that "boys will be boys," or to ask why a woman can't be more like a man? *Which* boy? Which woman, and which man?

These findings and patterns are awkward for those who want to argue that the sexes "naturally" segregate into different occupations and roles because of their different natures, or because of a slight advantage of one sex over the other, on average, on a particular trait. Job performance, paid or unpaid, depends on a suite of different skills, traits, interests, and values. People simply don't develop a career doing one thing really well, like identifying facial expressions of emotion, being sympathetic, or banging a fist on a boardroom table. What's more, for most jobs, there isn't one, single ideal combination of characteristics, skills, and motivations, but a range that could all fit the bill equally nicely. That's why not everyone at your level, in your role, in your occupation, is just like you. So if you want to, say, trot out the argument that women are just more psychologically suited to taking care of small children, you're committing yourself to the claim that women's hugely variable gender mosaics far more often match the many possible mosaics for caring for young children than do men's hugely variable gender mosaics. I don't say this kind of argument *can't* be successfully made. But I'd ask to see how you worked it out.

"THERE IS NO DOUBT that biology, via evolution and genetics, has made men and women significantly different."[93] So concludes Wolpert's book, in answer to the question posed by its UK and U.S. titles, *Why Can't a Woman Be More Like a Man?* and *Why Can't a Man Be More Like a Woman?* But as Valian astutely notes in *Nature*, "both titles suggest the retort: each can be."[94] And as we've also now seen, although there are certainly average sex differences, the very phrases "like a woman" and "like a man" make little sense at the level of brain and behaviour.

This isn't mere semantics or academic nitpicking. When young children and adults are asked to explain statements like "Boys have something called 'fibrinogen' in their blood," or "Boys are really good at a dance called 'quibbing,'" the kinds of explanations they come up with are different from the ones they create for statements like "This boy has fibrinogen in his blood," or "This boy is really good at quibbing." The *generic* statements more often trigger explanations grounded in assumptions that having fibrinogen in the blood, or being good at quibbing, or whatever it is, is *fundamental*—part of the true nature of being a male or a female. "I think it's a hormone that boys have because it is transcribed from male DNA," was one undergraduate's explanation of fibrinogen in the blood. "Boys are generally stronger than girls, and quibbing sounds like it requires some strength." By contrast, non-generic statements brought about relatively greater numbers of "non-essentialized" explanations that saw these characteristics as a one-off, even a problem, or due to external causes like practice or training. "A style of dance that the boy has practiced and is good at."[95]

When mosaics of mostly small average differences are carelessly squished into uni-dimensional generic claims—men are like *this*, women are more *that*—the natural inference is that we are talking about universal characteristics that are *"central, deep, stable, inherent*— in a word, 'essential.'"[96] When we say, think, or write statements like "Males are higher in competitiveness, dominance-seeking, and risk taking, while females are higher in nurturance,"[97] it's tempting for the mind to turn to the almighty T and the omnipotent Y of Casey's review—to *sex*—as a principal, powerful cause that sets us all on one of two divergent paths. But the overlapping, shifting, multidimensional, idiosyncratic mosaics formed by patterns of sex differences instead point to the combined and continuous action of many small causal influences.[98] Sex *doesn't* create male natures and female natures, and the next chapter turns to risk taking and competitiveness to complete the case.

CHAPTER 5

SKYDIVING WALLFLOWERS

My eldest son has long been irresistibly drawn to danger. At six months old he rolled across the entire expanse of the living room in order to more closely inspect the drill that his father—forgivably assuming that five yards was a safe distance to place a power tool from a baby who couldn't yet crawl—had put on the floor. On one memorable toddler playdate, within five minutes he had located the drawer of sharp kitchen knives that his little host Harry had failed to discover in his two years of life, and begun juggling with its contents. As a preschooler, whenever I took him to an indoor play centre—those brightly coloured monuments to the eradication of risk from childhood—my son would nonetheless routinely manage to imperil himself. At the age of ten, I left him happily engaged in the normally hazardless activity of assembling a cake batter, only to return five minutes later to discover him about to plunge a roaring hairdryer into the mixture. As he calmly explained, he had forgotten to melt the butter before adding it to the bowl, and was therefore trying to do so retroactively.

I admit that at times like these I have occasionally wondered why it was my lot in life to have a child so blasé about risk, and whether ultimately this will prove to be a blessing or a curse. On optimistic days I imagine him reaping enormous benefits: the invention of a time machine, say, after decades of dangerous experimentation. But in darker moments, I foresee much bleaker fates featuring mortuary lockers. While proponents of the Testosterone Rex perspective obviously don't share this fascination with my firstborn and his future, they do, as we saw in the Introduction, have a strong interest in the idea of risk taking as an inherently male trait. They would regard each of my son's perilous follies as successful manifestations of evolutionary pressures: a pitiful consolation, I can assure you, when you are trimming singed hair from your child's bangs and hoping the other guests at the barbecue won't ask too many questions. Economists Moshe Hoffman and Erez Yoeli recently spooled out the familiar chain of assumptions in the *Rady Business Journal*:

> When males take on extra risk in foraging for food, ousting rivals, and fighting over territory, they are rewarded with dozens, even hundreds of mates, and many, many babies. A worthwhile gamble! Not so for the females.[1]

Dozens? *Hundreds?* Sure—if you're a red deer, or the leader of an ancient Mongol empire. While Hoffman and Yoeli's arguments mostly refer to the "fighting" part of Darwin's sexual selection theory (intrasexual selection), other researchers suggest that risk taking also adds to men's appeal as a mate; the "charming" part of Darwin's subtheory (or intersexual selection). As psychologists Michael Baker Jr. and Jon Maner explain:

> Among men, risky behaviors have potential for displaying to potential mates characteristics such as social dominance,

> confidence, ambition, skill and mental acuity, all of which are highly desired by women seeking a romantic partner.[2]

But for women, there are no such benefits to be gained from taking risks. This is because—the authors seem to try to put it as tactfully as they can—"men tend to desire women with characteristics that signal high reproductive capacity (e.g., youth) rather than characteristics that might be signaled by risk-taking."[3] In other words, so long as the hair is glossy, the skin smooth, and the hip-to-waist ratio pleasing, then a cringingly low sense of self-worth, apathy, incompetence, and stupidity are relative trifles, more easily overlooked from the male perspective.

Having drawn on a vintage version of sexual selection to claim an evolutionary imperative for male risk taking, the next obvious step is to argue that this is a major contributor to persistent sex inequalities, helping to explain why fame, fortune, and corner offices are disproportionately acquired by men. Hoffman and Yoeli, for instance, argue that

> stocks have higher average returns than bonds, and competitive jobs can be quite lucrative. These rewards make gender differences in risk preferences one of the pre-eminent causes of gender difference in the labor market.[4]

The reference to competitive jobs points to a related explanation for occupational inequalities also much in vogue within the economics community: competition. Competition also involves risk taking since outcomes are uncertain, and the possible gains have to be weighed against the costs of taking part and defeat.[5] Thus:

> Over the past decade, economists have become increasingly interested in investigating whether gender differences in competitiveness may help explain why labor market differences persist. If women are more reluctant to compete,

then they may be less likely to seek promotions or to enter male-dominated and competitive fields.[6]

This leaves some mysteries to be explained—like young British women's considerable interest in competing for a place in the highly competitive, now slightly female-dominated, undergraduate courses of medicine and dentistry.[7] But even setting such issues aside, unravelling strand by strand this popular account of risk taking as an essential masculine trait reveals that just about every assumption on which it is based is wrong.

ALTHOUGH THE INSIGHT WILL probably fail to bring excitement to your next trip to the supermarket, there is an element of uncertainty to everything we do. Risk taking, in everyday understanding, is an action that potentially enables us to achieve a desired goal or benefit, but that also brings the possibility that we will fail, or something will go wrong. As a consequence, we may lose out on something we had, or could have had for sure (the kids' college fund, an unblemished reputation, the steady income from a government bond, a left arm) or, despite costly efforts, fail to gain something we hoped for (a date, a bulging pension fund, a prestigious promotion, a gold medal, the best-selling feminist science book the world has ever known). It's long been assumed that the propensity for risk taking is a stable personality trait—that is, a particular individual will consistently tend to seek, or avoid, risks in every realm of life. Indeed, for many years psychologists used measures of risk taking that added up a person's willingness to take risks in several different domains (like health, investments, and career) to yield a single risk-taking score.[8] Many economists, meanwhile, study risk taking by presenting participants with series of carefully designed lottery tasks, in which people choose between, say, $5 for sure or an 80 per cent chance of $10. The assumption is apparently that economists can infer "the" risk-taking profile of a person from his or her selections.[9]

The long-held belief that everyone can be neatly located at a point on a single continuum between "risk taker" and "risk avoider" fits nicely with the expectation for "a taste for competitive risk taking to be an evolved aspect of masculine psychology as a result of sexual selection."[10] Males, according to this view, are clustered mostly over on the risk-taking side, women on the careful end.

However, for decades there have been indications that risk taking isn't a one-dimensional personality trait: instead, there are "insurance-buying gamblers" and "skydiving wallflowers," as one group of researchers put it.[11] An early study of more than five hundred business executives, for example, looked at their preferences across a variety of risky choices, like business and personal investments, complex financial choice dilemmas, the amount of their own wealth held in risky assets, as well as non-financial risks. Clearly, if risk taking were a stable personality trait, then someone who tended to take risks in one area of decision making would also tend to report being a risk taker in the other domains. Yet this simply wasn't so. Knowing the riskiness of an executive's personal wealth strategy, for instance, told you nothing about how he'd behave in a business investment context.[12]

To investigate this surprising pattern more closely, Columbia University's Elke Weber and colleagues asked several hundred U.S. undergraduates how likely they would be to take risks in six different domains: gambling, financial, health, recreational, social, and ethical decisions. Again, a person's risk-taking propensity didn't follow any kind of consistent pattern across the different domains—that is, the person who would happily blow a week's wages at the races was no more likely to leap from a bridge attached to a rubber cord, invest in speculative stock, ask their boss for a raise, have unprotected sex, or steal an additional TV cable connection, than was someone who'd as soon flush dollar bills down the toilet as put them on a horse.[13] Researchers drew the same conclusion a few years later in a study that deliberately recruited people on the basis of their affinity to a particular kind of risk: like skydivers, smokers, casino gamblers, and members of stock-trading clubs. Once again, risk taking in one domain didn't extend to others. So gamblers, say, unsurprisingly stood out as the most risk taking when it came to questions about betting. But they were no more risk taking than the other groups, including even a group of health-risk-averse gym members, when it came to questions about recreational or investment risks.[14]

To see the problem this creates for the idea of risk taking as an essential masculine trait, ask yourself which group are the "real" men, or show a properly evolved masculine psychology: the skydivers, or the traders? That we expect Testosterone Rex to create an all-around risk taker is implicit in Hoffman and Yoeli's remark that men are "more likely [than women] to die in a car accident while speeding in the Ferrari they bought with their stock-market earnings." But as we've just seen, the reckless Ferrari driver might well prefer bonds to stocks. (That hypothetical ass probably inherited his wealth.) The pure, unadulterated daredevil no doubt exists, but such individuals are statistical exceptions to the general rule that people are fascinatingly idiosyncratic and multifaceted when it comes to risk.

So what makes someone eager to take risks in one domain, but reluctant in another? It turns out to be risk takers' less negative perception of the risks and more positive perception of the benefits, Weber and colleagues found.[15] The study of the skydivers, gamblers, smokers, and stock traders came to a similar conclusion. The risk takers in this study didn't like risk per se any more than did the risk-averse gym members. Rather, they perceived greater benefits in their particular pocket of risk taking, and this explained why they took risks that others avoided, and took one kind of risk rather than another. Similarly, *contra* common lore, entrepreneurs don't have a more indulgent, risk-loving attitude than do others towards the possibility of losing large sums of money; rather, they have greater confidence that everything will work out just fine.[16]

In fact, people are generally mildly risk averse.[17] This may at first seem to defy belief. However, *Chancing It* author Ralph Keyes drew exactly the same conclusion, based on extensive interviews about risk with people from many walks of life. One of his interviewees was the wire-walker Philippe Petit, famous for the remarkable feat of walking a wire strung a quarter of a mile high between the Twin Towers. However, Petit emphatically described himself to Keyes as "absolutely the contrary of a daredevil," adamantly declaring that "in no way, shape or form did he consider himself to be a taker of risks."[18] I was reminded of this remark at an extravagant magic show I saw recently with my children. In the melodramatic final act, an escape artist was handcuffed, then lowered upside down and padlocked into a snugly fitting glass tank of cold water, armed with only a single bobby pin with which to effect his escape. As we complacently watched the proceedings from the comfort of our seats, the master of ceremonies emphasized the extreme danger of the situation. Yet clearly, the theatre would have been heaving with the rush of parents clamping hands over children's eyes if there had been even a modest possibility that an afternoon's treat at the theatre would include witnessing a man drown on stage. By way of a less spectacular example of the principle that risk is in the eye of the beholder, my father, sister, and I can reliably evoke horror and fear in dinner guests with our blasé attitudes towards health and safety in food storage and preparation. Questions like *Which is your board for chopping meat?* from helpful guests are invariably met with blank, uncomprehending stares. But none of us Fines think we are dicing with danger when we dice vegetables on a board smeared with raw chicken. We simply have a profound (and so far fully justified) confidence that festering microbes are no match for the notoriously robust Fine constitution.

The critical point here is that "the risk in a given situation is inherently subjective, varying from one individual to the next."[19] It's simply not possible to assess the "objective" characteristics of a risky situation, then infer a person's appetite for risk from the decision she or he makes. Again, this resonates with the conclusion Keyes draws. "Repeatedly," he writes, "I've discovered that those who are apparently taking big risks turned out on closer examination to be risking little; little of value, that is." In a rhetorical question that pointedly emphasizes the subjectivity of the potential losses and gains at stake, he asks: "If you risk a life you don't value, have you taken a risk?"[20]

The importance of subjectivity in the perception of risks and benefits for humanity's colourful diversity of risk taking turns out to be equally crucial for understanding sex differences. Contrary to what many might assume, women and men have similar risk attitudes, Weber and colleagues found. For the same subjectively perceived risk and benefit, they are equally likely to tempt fate. When men and women *do* diverge in risk-taking propensity, it is because they perceive the risks and benefits differently.[21] So are men inherently constituted to perceive risks more positively, making them more inclined to take them? A closer look at the actual pattern of sex differences in risk taking reveals important nuances that make this unworkable as an explanation.

A good starting point is a large meta-analysis that collated studies of female/male differences in risk taking across a variety of domains (like hypothetical choices, drinking, drugs, sexual activity, and driving), and across five different age groups from middle childhood to adulthood.[22] This analysis did indeed lead to the conclusion that males are more risk taking than females, on average. But about half of the differences were very modest, and in 20 per cent of cases they were even in the "wrong" direction (that is, there was greater *female* risk taking). The meta-analysis also revealed changeable patterns of difference depending on the age group and the kind of risk. For example, studies of eighteen- to twenty-one-year olds found that males were a little more likely, on average, to report drinking and drug taking, and risky sexual activities. But for older adults, this sex difference was almost exactly reversed. Nor was there any obvious pattern in the effect of age on sex differences. This is surprising: if risk taking evolved to increase reproductive success, you'd presumably expect an especially clear divergence of the sexes following pubescence. The traditional view of risk taking as a masculine trait therefore requires revision, the researchers concluded, in the face of evidence that "risk taking . . . does not seem to manifest itself in a simple or constant way across ages or contexts."[23]

That only some domains favour male risk taking leads to the important point that, given an imperfect world in which people can and do die by falling out of bed or accidentally swallowing toothpicks, researchers have to make decisions about the kinds of risks they decide to investigate. With risk taking intimately linked with masculinity in our minds (it's no mere coincidence, surely, that business jargon for a bold vision is a *big, hairy, audacious* goal), it's easy to fail to notice what *doesn't* tend to make it onto the questionnaires. What about the surprisingly dangerous sport of cheerleading, or galloping across a field on a horse, or Bingo? As University of Massachusetts Boston economist Julie Nelson notes, although women routinely take risks, these often seem to slip under the research radar.[24] For example, with divorce rates hovering close to 50 per cent, being the one to quit or scale back your job when children arrive is a significant economic risk. Going on a date can end in sexual assault. Leaving a marriage is financially, socially, and emotionally risky. In the United States, being pregnant is about twenty times more likely to result in death than is a skydive.[25] And simply slipping on a pair of high heels in the morning increases the risk of chronic pain, irreversible leg tendon damage, osteoarthritis of the knee, plantar fasciitis, sciatica,[26] and (if you will forgive just one more technical term) the painful and embarrassing condition of fallingflatonyourfaceitis.[27] None of which is to say that existing assessments of sex differences in risk taking aren't informative, interesting, and valid. However, they also reflect implicitly gendered assumptions about what risk taking is. The reported gender gap in risk taking would almost certainly narrow if researchers' questionnaires started to include more items like *How likely is it that you would bake an impressive but difficult soufflé for an important dinner party, risk misogynist backlash by writing a feminist opinion piece, or train for a lucrative career in which there's a high probability of sex-based discrimination and harassment?*

Indeed, there are already documented exceptions to the notion of risk taking as a masculine trait. A number of studies have found that women are at least as willing as men to take social risks (like admitting that their tastes are different from those of their friends, or disagreeing with their father on a major issue).[28] Women were also found to be more likely than men to report that they would take risks in situations in which there was a small chance of benefit for a small fixed cost (such as trying to sell an already-written screenplay to a Hollywood studio, or calling a radio station running a promotion in which the twelfth caller receives money).[29] So why do perceptions of risks and benefits apparently differ between the sexes in some realms but not others? One obvious answer is that some activities—like unprotected sex or excessive drinking—may actually *be* objectively more risky for females. Risk researchers have also found that both knowledge and familiarity in a particular domain reduces perceptions of risk.[30] Plausibly, men may tend to be relatively more knowledgeable or familiar with some of the risky activities that tend to feature in surveys (like sports betting, financial investments, and motorcycle riding).

The point is that an "unruly amalgam of things" underlies choices, as Harvard University legal scholar Cass Sunstein puts it: "aspirations, tastes, physical states, responses to existing roles and norms, values, judgments, emotions, drives, beliefs, whims."[31] And so, we're not only sensitive to the material benefits and costs when we make choices, Sunstein argues, but also to the less tangible effects a particular choice could have on self-concept and reputation. In a gendered world, these impacts are inevitably different for females and males. (Recall, for example, the different anticipated sexual pleasure and cost to reputation of a casual sexual encounter revealed by Terri Conley and colleagues' research, described in Chapter 2.) A striking demonstration comes from research investigating the risks people perceive from technological, lifestyle, and environmental hazards (like nuclear power, smoking, and ozone depletion). These studies routinely find that women perceive higher risks to themselves, family, and society from such hazards.[32] For instance, James Flynn and colleagues surveyed more than fifteen hundred U.S. households and found that women on average perceived higher risks across the board.[33] The Testosterone Rex explanation of this would be that women, as the nurturers of precious offspring, have evolved to be more cautious about threats to physical health. However, Flynn and colleagues then subdivided the sample by ethnicity as well as sex, and discovered that one subgroup stood out from all the rest. Society seemed a significantly safer place to white males than it did to all other groups, including nonwhite men. What on first inspection seemed like a sex difference was actually a difference between white males and everyone else.

Flynn and colleagues then established that it was a particular subset of white males who were particularly cavalier about risks: those who, in response to the social justice movement's currently fashionable suggestion to "check your privilege," would take significantly longer than others to complete the task. These men were well educated, rich, and politically conservative, as well as more trusting of institutions and authorities, and opposed to a "power to the people" view of the world. A number of studies have now replicated this so-called "white male effect" with other large U.S. samples,[34] and the research points to it being "not so much a 'white male effect' as a 'white hierarchical and individualistic male effect.'"[35] While I could tell you the kinds of statements these men tend to agree with (*We have gone too far in pushing equal rights in this country. . . . A lot of problems in our society today come from the decline in the traditional family*), and disagree with (*Sometimes government needs to make laws that keep people from hurting themselves. . . . It's society's responsibility to make sure everyone's basic needs are met . . .*) it might be easier and quicker to simply picture Glenn Beck.

Interestingly, a recent study conducted in the more socially egalitarian and gender-equal Sweden failed to find the "white male effect." This national survey of nearly fifteen hundred households found that, all else being equal—and in stark contrast with the U.S.

data—Swedish men and women had very similar perceptions of lifestyle, environmental, technological, health, and social risks.[36] The survey found instead just a "white effect," with people from foreign backgrounds, who are subject to social disenfranchisement and discrimination, perceiving risks as higher than did native Swedes.

In trying to understand how social place and identity can affect risk perception in such a marked way, it's helpful to know that people often use their feelings as a guide to the risk-benefit trade-off. The more positively we feel about something—whether it's unpasteurized French cheese, vaccinations, or abortion—the more we tend to minimize the risks and play up the benefits. Conversely, if we feel antipathy towards an activity or hazard, we "tend to judge the opposite—high risk and low benefit."[37] Political worldview is a potent source of strong emotions about risky hazards, and it may be that people perceive risk in ways that protect their social identities, roles, and status:

> Perhaps white males see less risk in the world because they create, manage, control, and benefit from so much of it. Perhaps women and non-white men see the world as more dangerous because in many ways they are more vulnerable, because they benefit less from many of its technologies and institutions, and because they have less power and control.[38]

This point was neatly demonstrated by some statistical fun, inspired by Nelson's insight that we tend to *think risk, think male.* Yale Law School academic Dan Kahan showed that, when asked about the risks to human health, safety, or prosperity arising from high tax rates for business, now it was the women's and minority men's turn to be sanguine. This, he notes, beautifully illustrates Nelson's point:

> It confirms that men are more risk tolerant than women *only* if some unexamined premise about what counts as a "risk"

> excludes from assessment the sorts of things that scare the pants off of white men (or at least hierarchical, individualistic ones).[39]

The white male effect in the United States, viewed alongside the similar risk perceptions of native Swedish men and women, suggests that it can at least sometimes be the different social place, identities, and experiences of men and women in the world, rather than some enduring dissimilarity of biology, that underlie sex differences in risk perception. This is a vital point since, as we've seen, it is these subjective perceptions that underlie sex differences in risk taking. The idea that women have evolved to be biologically predisposed to perceive greater risks to health is intuitively plausible, but appears to be simply wrong. As the researchers who first identified the white male effect point out: "Biological factors should apply to nonwhite men and women as well as to white people."[40]

No less importantly, social identities come in a package that includes social norms. These norms, as Sunstein has emphasized, play a crucial role in our decision making. Indeed, psychologists Catherine Rawn and Kathleen Vohs have compiled a convincing case that people sometimes overcome strong preferences to *avoid* risky but socially expected behaviours (such as drinking, drugs, sex, or violence) in order to stay "in" with others.[41] Gender, of course, is a rich source of norms that apply differently to males and females, with some behaviours more strongly expected of one sex, and others more strongly censured.[42] For example, there is a stronger expectation of women to "be nice" than there is of men. When women violate this norm in a workplace setting (by behaving in domineering ways or negotiating for better remuneration and conditions, for instance) they encounter backlash from others, who become less willing to work with them, and like them less.[43] This means that statements that men "are more likely to bargain aggressively for their starting salaries"[44] need some unpacking. If so, is it really because women are intrinsically risk averse, or care less about money? Or is it because there is a violation of feminine norms involved in bargaining aggressively in one's own self-interest, and so women quite accurately intuit a less favourable balance of benefits and risks from doing so?

On the first point, research has found that a sex difference in negotiating for bigger pay (in a lab task) can be eliminated simply by framing exactly the same behaviour in a way that's more in keeping with feminine norms of politeness: "asking," rather than negotiating. As the authors point out, "the term *negotiation* is not gender-neutral."[45] And regarding the second point, does violating those norms yield the same benefits? One study found that although top-flight female MBA students were just as likely as their male counterparts to negotiate their initial post-course salary, the financial pay-off for them was less.[46] It's not hard to imagine those women being less likely to negotiate in the future, but because of the anticipation of less reward rather than an evolved disinclination to take risks. Exeter University psychologist Michelle Ryan surveyed more than eight hundred managers at a major consultancy firm, and found that women on average were less willing than men to make sacrifices for their career, and to take career risks in order to get ahead. Closer examination revealed that this was because women tended to perceive less benefit in taking risks and making sacrifices. But this was not because they were simply less ambitious. Rather, they had lower expectations of success, fewer role models, less support, and less confidence that their organization was a meritocracy.[47]

In many domains, gender norms tend to favour *male* risk taking, which is a norm of masculinity[48] and seen as a more important trait for men than for women.[49] This means that, in addition to material gains, taking risks may often bring greater reputational benefits or smaller costs: women in counter-stereotypical leadership roles are judged more harshly than men when risky decisions don't work out.[50] Underscoring the importance of Sunstein's unruly amalgam, both men and women seem to be responsive to cultural information about how their risk-taking behaviour will be perceived by others. In one study, for instance, single men presented with a newspaper article claiming that women found risk taking unattractive in partners subsequently made less risky choices in a lab task administered by a female experimenter (compared with men who'd read a stereotype-consistent article).[51] Or consider a recent study of young Chinese women and men, who played a risk-taking game either privately or while being were observed by an attractive person of the other sex. In China, the authors argue, the ideal of womanhood strongly precludes risk taking, valuing instead those who are "timid, reserved, shy, obedient, unassertive, humble, attentive, respectful, and, above all, chaste."[52] In contrast with this feminine ideal, Chinese women were every bit as risk taking as the men when unobserved. But in line with gender norms, men increased their risk taking when supposedly observed by an attractive other-sex observer, while women decreased it.

Some, of course, might argue that an asymmetry in gender norms for risk taking is nonetheless inescapable, thanks to the evolutionary advantages to males of risk taking, from which female partners then benefit. As we saw in the first part of the book, this argument requires overlooking the reproductive advantage to *males* of selecting a mate who can hold her own in the reproductive stakes. But there's also a more directly devastating problem: women *aren't* drawn to risk takers. Gambling, ethical risk taking, and health risks are all viewed as unattractive in potential mates, and even financial risk taking is touch and go.[53] Social risk taking, by contrast, *is* alluring in a potential partner (things like being willing to defend an unpopular position at a social event). But as you'll recall, women are just as likely as men to take social risks. And while physical risk taking is also viewed positively, especially in people being considered for a short-term relationship, this is only if the level of risk is perceived to be low. People desire "neither daredevils nor wimps"[54] and, surprisingly, "the less risky an [activity] is perceived to be, the more attractive it is."[55] This is all a far cry from the assumption of female glorification of male risk taking. But most importantly, this pattern of preferences is no less true of men than it is of women: heterosexual men are generally no less attracted to physical and social risk takers than are women.[56]

This is a problem for the Testosterone Rex view. Some research teams have acknowledged this minor devastation of a pet hypothesis with good grace, accepting that the proposal that men have evolved to display risky behaviour in order to attract females "does not account for the observed similarities between men and women" and "offers little explanation . . . for men's preference for risk-takers."[57] Summarizing this "picture of overall similarity between the sexes," Andreas Wilke of the Max Planck Institute for Human Development and colleagues suggest instead

> that men and women learn to value the same traits for non-adaptive reasons (e.g., a cultural norm) or that the same sort of risk taking might (at least in societies with male investment comparable to female levels) be a reliable cue to quality for both sexes.[58]

In other words, maybe there isn't anything so special about male risk taking, after all.

Nor, it turns out, are females quite as disinclined to take the risk of going head-to-head with others in a competitive context as they are often assumed to be. The Testosterone Rex view of competition inspired by those hypothetical one hundred babies in a year from a hundred different women, leads to a simple "men are more competitive" prediction:

> Relative to females, male reproductive success is affected more by their ability to obtain mates. Males may compete directly for mates or they may compete for resources, territory or status, all or which serve to increase their mating opportunities. . . . Consequently, the preference to compete should be more pronounced in men compared to women.[59]

But one of the few psychological studies to look at the frequency of competitive behaviour in the real world—two diary studies of UK students by Elizabeth Cashdan—found nothing of the kind.[60] Women and men reported similar rates of competing and the sexes were also remarkably similar in terms of how often they vied with others within particular domains. They were equally competitive with regard to their studies and work (arguably the best route to future economic resources for students), and status (which ranked rather low in both studies). The only arena in which men were more likely to compete than women was in sports, while women were more likely to compete over "looking good," neither of which seem key to understanding occupational sex inequality.

Nor have economists' more tightly controlled studies invariably found that males are more competitive. In this discipline, the standard approach is to give participants some kind of bloodless task (the somewhat masculine-flavoured staples include adding three-digit numbers or throwing balls into buckets). Then having tried their hand at the task, each participant is given the choice to either earn a modest "piece rate" for each success, *or* a much more generous sum per success, so long as they beat a randomly selected opponent. Whether or not you see sex differences turns out to depend on what you are asking people to compete *at*, and also *which* males and females you are asking. When researchers use more neutral or "feminine" competitive contexts—like dancing, verbal ability, fashion knowledge, or a stereotypically feminine job (like "administrative assistant" versus "sports news assistant"), they often find that females are equally, or sometimes even more, likely to compete.[61] The cultural background of participants also has a significant impact on whether or not sex differences are seen; interestingly, coming from a homeland with lesser economic development seems to be associated with greater female competition.[62] Thus, Colombian, Han Chinese, and Armenian girls and Beijing women are as competitive as their male counterparts, even in the kinds of tasks in which, in developed Western cultures, greater male competitiveness is the typical finding.[63]

Most striking of all, while men from the patriarchal Maasai society in Tanzania were more willing than women to compete against others to try to earn money by successfully throwing tennis balls into a bucket, exactly the opposite was seen in the matrilineal Khasi society in India.[64] Moreover, among the children from these societies, only in the patriarchal Maasai did boys become more competitive than girls in the post-pubescence years.[65]

That biology clearly doesn't determine that males should be more competitive than females makes it all the more concerning that, among Austrian children of three to four years of age, boys are already more willing to compete in a running competition than are girls (even though girls can run just as fast). At this age, girls are as eager to compete in a more "feminine" manual sorting task (at which they are slightly superior); but within a few years, even here, boys are more competitive.[66] What messages are children receiving in developed Western countries that, compared with children in some other societies, separates girls' and boys' inclination to compete from such an early age?

WRITING IN THE *FINANCIAL TIMES* about our supposedly admiring attitudes towards those who take big financial risks, columnist John Kay draws direct links to our Stone Age past, contrasting "prudent hunters" who peered about anxiously for dangerous animals and "stayed at home when it was too dangerous," with more courageous hunters who "chose not to buy or exercise these options" and therefore "took more risks and caught more prey."[67] Lest there be any doubt in readers' minds as to which sex it was that won admiration with their daring, Kay then rhetorically asks, "Were the young women of the tribe more impressed when the cautious described their uneventful days, or when the bold recalled their heroic escape from danger?" For some reason, Kay omits asking readers to consider how the women would have appreciated the conversation of hunters whose throats had been ripped out by wild animals.

We are now a long way from the bundle of assumptions packed into this familiar vignette. Risk taking is not a stable personality trait, allowing us to assume that the person who would willingly take the physical risks of hunting (or white-water rafting or skydiving) would be a fearless CEO or trader. Nor is risk taking something that only men do, or that only women are drawn to in a potential partner. Meanwhile, the growing evidence that females *do* compete at equal rates to men when the nature of the task seems to authorize it, and that girls and women from populations further afield than the typical Western samples are no less eager than males to compete, undermines assumptions that this is an "essential" sex difference.

What does this mean for assumptions of "testosterone-fuelled" male risk taking? From the old understanding of risk taking as a masculine trait, sex differences in testosterone is an intuitive, obvious, and common explanation. But as the last chapter argued—and the next chapter further reinforces—the shape and pattern of sex differences defies explanations in terms of a single, powerful cause that splits the sexes.

As I was editing this chapter, a survey of more than thirty-five hundred Australian surgeons revealed a culture rife with bullying, discrimination, and sexual harassment, against women especially (although men weren't untouched either). To give you a flavour of professional life as a woman in this field, female trainees and junior surgeons "reported feeling obliged to give their supervisors sexual favours to keep their jobs"; endured flagrantly illegal hostility towards the notion of combining career with motherhood; contended with "boys' clubs"; and experienced entrenched sexism at all levels and "a culture of fear and reprisal, with known bullies in senior positions seen as untouchable."[68] I came back to this chapter on the very day that news broke in the state of Victoria, Australia, where I live, of a Victorian Equal Opportunity and Human Rights Commission report revealing that sexual discrimination and harassment is also shockingly prevalent in the Victorian Police, which unlawfully failed to provide an equal and safe working environment.[69]

I understand that attempts to identify the psychological factors that underlie sex inequalities in the workplace are well-meaning. And, of course, we shouldn't shy away from naming (supposedly) politically unpalatable causes of those inequalities. But when you consider the women who enter and persist in highly competitive and risky occupations like surgery and policing—despite the odds stacked against them by largely unfettered sex discrimination and harassment—casual scholarly suggestions that women are relatively few in number, particularly in the higher echelons, because they're less geared to compete in the workplace, start to seem almost offensive.

Testosterone Rex implicitly blames women for their lower salary and status, distracting attention away from the "unruly amalgam" of gendered influences—the norms, beliefs, rewards, inequalities, experiences, and, let's not forget, punishment by those who seek to protect their turf from lower-status outsiders—that unevenly tip the cost-benefit scales.

CHAPTER 6

THE HORMONAL ESSENCE OF THE T-REX?

Adult brains are like celestial objects or continents—more dynamic and plastic than most scientists used to imagine.

—ELIZABETH ADKINS-REGAN, *Hormones and Animal Social Behavior*[1]

SOMETIMES THESE DAYS I'M INTRODUCED TO PEOPLE AS AN academic who wrote a book about how the brains of women and men aren't that different. Disappointingly, the wide range of reactions to this brief biography has yet to include *You must be Cordelia Fine! Would you sign this copy of your book that I carry around with me?* Instead, people often shoot me a startled look, and then ask whether I'd also deny that there are other basic physiological differences between the sexes. Whenever this happens, I'm always tempted to fix my interrogator in the grip of a steely gaze and pronounce briskly, "Certainly! Testes are merely a social construction," then see how the conversation flows from there.

Needless to say, this would be especially mischievous given the presumed role of the testes as the biological wellspring of the hormonal essence of masculinity—that steroid tsunami that destroys all hopes of sex equality. As Wayne State University law professor Kingsley Browne recently put it:

> Despite the frequent assertion that the gaps that favor men (although not those that favor women) are results of invidious social forces, the truth seems to be somewhat more basic. If the various workplace and non-workplace gaps could be distilled down to a single word, that word would not be "discrimination" but "testosterone."[2]

In much the same way, economists who suggest that inherent sex differences in risk taking play a major role in economic and occupational inequalities sometimes finger testosterone as the biological culprit.[3] And according to neuroscientist Joe Herbert, author of *Testosterone: Sex, Power, and the Will to Win*, "the testis is the source of most of what we term masculinity."[4] This is apparently because the testosterone it produces "prepares males for the rigorous and competitive events of reproduction." Thus, he writes:

> Testosterone has to do a great many things: it must influence physique; act on the brain; and inflame sexuality. But this hormone also makes males enjoy taking risks, resorting readily to competitiveness and aggression to obtain what they need, seeking domination over other males, resenting and repelling invasion of their territory."[5]

That's a big portfolio.

Science writer and behavioural endocrinologist Richard Francis coined the term "Testosterone Rex" to poke fun at the mistaken conception of testosterone as a "super-actor"—the "plenipotent executor of selection's demands" that simply "takes care of everything."[6] Certainly, if the problem to be taken care of is how to create two kinds of individual, then testosterone as a plenipotent super-actor offers a neat and obvious solution. While scientific views on T's role in social behaviour vary around the edges, they generally point to a link with competition as key.[7] Most obviously, this refers to competition to acquire or defend social status, material resources, and sexual opportunities. However, it should also probably include a facet of parenting—protection of that most precious resource, offspring, argues University of Michigan social neuroendocrinologist Sari van Anders. Low T, by contrast, is linked with nurturance.[8] So according to a T-Rex view, high-T individuals cluster at the competitive end of the continuum with the other aggressive, sexually inflamed risk takers, while low-T characters huddle at the duller but safer and more caring opposite pole.

Consider, for example, a cichlid fish known as *Haplochromis burtoni* that comes from the lakes of East Africa.[9] In this species, only a small number of males secure a breeding territory, and they are not discreet about their privileged social status. In contrast to their drably beige non-territorial counterparts, territorial males sport bold splashes of red and orange, and intimidating black eye stripes. The typical day for a territorial male involves a busy schedule of unreconstructed masculinity: fighting off intruders, risking predation in order to woo a female into his territory, then, having inseminated her by ejaculating into her mouth, immediately setting off in pursuit of a new female. Add to this the fact that territorial males boast significantly larger testes and have higher circulating levels of testosterone than submissive non-territorial males, and a T-Rex view of the situation seems almost irresistible. These high-T fish are kings indeed, presumably thanks to the effects of all that testosterone on their bodies, brain, and behaviour. With a large dose of artistic license, we might even imagine the reaction were a group of feminist cichlid fish to start agitating for greater territorial equality between the sexes. It's not discrimination, the feminist fish would be told, in tones of regret almost thick enough to hide the condescension, but testosterone.

But even in the cichlid fish, testosterone isn't the omnipotent player it at first seems to be. If it were, then castrating a territorial fish would be a guaranteed method of bringing about his social downfall.

Yet it isn't. When a castrated territorial fish is put in a tank with an intact non-territorial male of a similar size, the castrated male continues to dominate (although less aggressively). Despite his flatlined T levels, the status quo persists.[10] If you want to bring down a territorial male, no radical surgical operations are required. Instead, simply put him in a tank with a larger territorial male fish. Within a few days, the smaller male will lose his bold colours, neurons in a region of the brain involved in gonadal activity will reduce in size, and his testes will also correspondingly shrink. Exactly the opposite happens when a previously submissive, non-territorial male is experimentally manoeuvred into envied territorial status (by moving him into a new community with only females and smaller males): the neurons that direct gonadal growth expand, and his testes—the primary source of testosterone production—enlarge.[11] In other words, the T-Rex scenario places the chain of events precisely the wrong way around. As Francis and his colleagues, who carried out these studies, conclude: "Social events regulate gonadal events."[12] Or to put it another way, just in case the significance of this sailed past unnoticed, cichlid testes are a social construction.

In fact, even without looking at any data from behavioural endocrinology, suspicions about the T-Rex story should already be aroused. Recall the major conceptual and empirical shifts in sexual selection theory and research we met in the first part of the book. These have left the old assumptions that competition for mates, status, and resources are exclusively male pursuits in the fight for reproductive success gathering dust.[13] By way of example from a different species of fish, Sarah Blaffer Hrdy described long ago how female coho salmon compete ferociously for nests in which to bury their eggs. These head-to-heads have such serious reproductive consequences that a third of the time a defeated female's nest will be taken over and her eggs destroyed.[14] So why wouldn't some females also need a hormone to prepare *them* for the "rigorous and competitive events of reproduction"? As Cornell University neuroendocrinologist Elizabeth Adkins-Regan observes:

> Many females are very aggressive, sometimes more so than males, aggression among females is an important dyadic level process driving the spacing patterns and social systems of many animals, and in mammals the fitness consequences of rank in a dominance hierarchy are better established for females than for males.[15]

Already, then, we should be sceptical that, as a general rule, T serves to polarize the competitive behaviour of the sexes. At the very least, the situation needs to be taken on a species-by-species basis. And when we turn to ourselves in light of what we've learned in the last few chapters, we immediately encounter a problem. The T-Rex view would work fine if men were like *this* and women were like *that*. When we make generic statements like "men are competitive, women are caring," T differences seem like an obvious explanation. But can the T-Rex story explain the shape that sex differences *actually* take? How, for instance, does T-Rex make "boys be boys" when, as we saw in Chapter 4, there's no essential masculine profile that simultaneously unites a boy or man with most other males, and cleanly separates him from females? How does the T-Rex story deal with the fact that gendered behaviour doesn't, as was once thought, create a single dimension that runs from masculinity to femininity, or even two dimensions? Only when working from those simpler, outdated, one- or two-dimensional understandings of gender does it make sense to suppose that higher T could increase an individual's masculinity, and/or decrease their femininity. But this just doesn't work as an idea when masculinity and femininity are multidimensional, with most people possessing "a complicated array of masculine and feminine characteristics," as Joel puts it.[16] What particular attributes of masculinity should we expect a high-T man to show, or a low-T woman to lack? And in particular, how does T make males risk taking and competitive when, as we saw in the previous chapter, in some domains, contexts, and populations, female risk taking and competitiveness is equal to (or even surpasses) that of males?

Or, to repeat the awkward question that chapter posed, given that risk taking is domain specific—the physical daredevil may well be socially or financially risk averse—what kind of risk taker should we expect our high-T guy to be?

Fortunately, we don't have to answer difficult questions like these. This is because, in the evolution of scientific understanding of the relations between hormones and social behaviour, the notion of testosterone as the powerful hormonal essence of Testosterone Rex has not survived.

SPECULATION ABOUT TESTOSTERONE and behaviour has a long history. In his classic essay, *The Trouble with Testosterone*, celebrated neurobiologist and writer Robert Sapolsky hazards that "A dozen or so millennia ago, an adventurous soul managed to lop off a surly bull's testicles and thus invented behavioral endocrinology" (that is, the study of the relations between hormones and behaviour). This inadvertent experiment

> generated an influential finding—something or other comes out of the testes that helps to make males such aggressive pains in the ass.
>
> That something or other is testosterone.[17]

However, it wasn't until the mid-nineteenth century that the first formal experiments on testosterone-behaviour relations took place, in the busy hands of a German physiologist named Arnold Berthold.[18] Berthold's investigations began with the observation that when a cockerel is castrated, not only does its distinctively male comb go into retreat, but it also quits its roosterish lifestyle of fighting, mounting, and cock-a-doodle-doo-ing. Berthold then took the natural next step for an inquiring mind unbounded by squeamishness. He decided to see what happened when he either re-implanted the testes or, in other experiments (perhaps performed on days when he was in an especially macabre mood), when the testes were transplanted into the cockerel's stomach. Berthold's remarkable discovery was that both of these interventions restored the cockerel's cockiness. Since the newly located testes were no longer connected to the nervous system, Berthold was able to infer the action of something secreted into the bloodstream—a hormone. As we now know, testosterone and other androgens (the class of steroid hormone to which T belongs) are secreted into the bloodstream by the gonads (both testes and ovaries produce both androgens and oestrogens) and the adrenal glands.

The classic "remove-and-replace" experiments, of which there are now hundreds, established that testosterone has important effects on both the body (as on the wattle and comb, if you happen to be a rooster) and mating behaviour. Pointing to the same conclusions are the experiments of nature that take place when animals shift between life-history stages, for example, from youth to adulthood (or in some species, from a small size to a more imposing stature), or in and out of a breeding season. In our own species, of course, the gonads start to produce both androgens and oestrogens with renewed vigour (following the prenatal flurry) in pubescence, helping to bring about the development of secondary sexual characteristics. Some species of fish can even pull off the remarkable hormonal trick of changing sex when the opportunity (such as the death or removal of the dominant male in the group) arises.

This brings us to the important question of what hormones like T are *for*. In the first part of this book we met the idea that many animals only expend the biological costs of secondary sexual characteristics, and take the time, effort, and risk of courtship, when there's a good chance of mating and fertilization taking place. Hormones can help coordinate this, by synchronizing the necessary changes in body and behaviour. But T can also help coordinate individuals' behaviour on a shorter time scale. In a complicated and cruelly unpredictable world, it's simply not the case that successful arrival at a particular life stage means that a single way of being in the world will suffice. Hormones "help adjust behavior to circumstances and contexts," Adkins-Regan explains, "physical, social, and developmental."[19]

One way T can do this is easily overlooked—via its effect on the body. T alters the body in more or less spectacular ways, depending on the species, and these masculinized features can then evoke particular responses in others. We met an example of this in Chapter 4: mother rats, attracted to the higher levels of testosterone in male pups' urine, more intensively lick the anogenital region of their male offspring. This extra stimulation, we saw, ultimately contributes to sex differences in the brain and mating behaviour.[20] A less subtle example, bypassing the brain altogether, is the "sword" of the male swordfish that develops when T increases during sexual maturation. Females are attracted to the sword, and so the male's response to the female's sexual interest is therefore in a sense "caused" by T, but in a rather indirect way.[21] As for ourselves, there's a case to be made that the pervasive and comprehensive gender socialization that penetrates just about every aspect of human culture is just another example of the indirect effects of sex hormones—via their effects on the body that identify us as female or male—on behaviour.

But testosterone does also affect the brain directly.[22] In more lasting effects that take place at critical junctures in life—such as prenatally (in interaction with several other factors, as we saw in Chapter 4), in pubescence, or when spring is in the air—T helps to restructure neural pathways. T can also influence existing neural pathways in a more transient fashion (on the scale of minutes to weeks, depending on the mechanism), by either ramping up or down the electrical "excitability" of brain cells.[23] The intricacies of how it does this show just how much goes on that expressions like *it's the testosterone* overlook. In the fastest version of these short-term effects, T binds to the nerve cell membrane and, by altering chemical pathways, changes how readily a neuron fires.[24] However, the best-known route by which T affects the brain is via hormone receptors. T binds to an androgen receptor, and is then "escorted" into the nucleus of the nerve cell. There, its next step is to "tickle the genome."[25] Then, in combination with what are called "cofactors," a particular hormone-sensitive region of the gene is "turned on," altering protein and peptide production (or gene "expression"). Sometimes though, with the assistance of a biological catalyst called aromatase, T converts from a "male" androgen to a "female" oestrogen, then binds to an oestrogen receptor. (Yes, even the "sex hormones" defy the gender binary.) Alternatively, the oestrogen might not originate from testosterone, or even from the gonads, since it turns out that the brain can synthesize its own oestrogens from scratch.[26] Ultimately, this dance between steroid hormones and receptor can lead to a host of "behavior-impacting gene products," as Adkins-Regan puts it, from enzymes involved in producing steroids, steroid receptors, and neurotransmitters to proteins that help build and repair neurons:

> Through their intracellular receptors steroids alter neural activity now and in the future, alter their own production and reception and that of other steroids, and regulate some of the other neural signalling systems important for social behavior.[27]

In short, T certainly does stuff—important stuff. But now we get to the second reason for making you endure that dense last paragraph. Even though it barely begins to scratch the surface of the daunting complexities involved, it already makes clear that the amount of testosterone circulating in the bloodstream is just one part of a highly complicated system—the one that happens to be the easiest to measure.[28] The many other factors in the system—the cofactors, the conversion to oestrogen, how much aromatase is around to make that happen, the amount of oestrogen produced by the brain itself, the number and nature of androgen and oestrogen receptors, where they are located, their sensitivity—mean that the absolute testosterone level in the blood or saliva is likely to be an extremely crude guide to testosterone's effect on the brain.

This complexity may have made the preceding pages rough going, but it has a few useful consequences in the grander scheme of things. First of all, it means there's scope for evolution to have moulded this multilayered system according to each species' needs. T is ubiquitous among sexually reproducing species, but by tinkering with other factors, it's possible for "the degree of association between hormones and behavior to vary."[29] And in fact, evolution seems to have done exactly that. The hypothetical neuroendocrinologist who clung to the Bateman-inspired hope that T will affect animals in similar ways across the sexually reproducing animal kingdom would be doomed to a life of repeated disappointments.[30] This, in turn, means that just because testosterone has a particular effect on the behaviour of, say, elephant seals or bulls, doesn't guarantee the same consequences in humans.

The complexity also helps to make the following problem less bewildering. How do humans achieve the feat of turning something rather large (average sex differences in circulating testosterone levels) into something usually rather small (average sex differences in behaviour)? No sex difference in basic behaviour comes close to the divergence between the sexes in circulating testosterone, for which there's only about 10–15 per cent overlap between men's and women's levels.[31] Potentially, this puzzle is solved by the important principle we met in Chapter 4: that sex effects in the brain don't always serve to create different behaviour. Sometimes instead, one sex effect counteracts or compensates for another, enabling *similarity* of behaviour, despite dissimilarity of biology.[32] Combine this principle with the considerable room for manoeuvre in the journey between T in the bloodstream and its action on the brain, and it becomes clear that there are potential ways for the relative testosterone-y-ness of males to be ramped down. One researcher, for instance, suggests that male exposure to the testosterone surge *in utero* somehow *desensitizes* the brain to testosterone's effects later in life.[33] This would be a smart way, maybe achieved through sex differences in neural sensitivity,[34] of enabling males to tolerate the higher levels of testosterone their bodies need to develop and maintain male secondary sexual characteristics, without having an excessively large effect on behaviour.[35]

This brings us to another important point. T is often thought of as a "male" hormone, the assumption presumably being that only males have enough for it to be of psychological significance. When, after all, was the last time you heard someone despairingly say *"It's the testosterone"* of a *woman's* behaviour? Unless her transgression was to grow a beard, probably never. This popular *T = male* perception is both reflected in, and reinforced by, the much greater research attention on males than females.[36] But as van Anders wryly asks: "What does its natural occurrence do in *females,* then?"[37] As she points out, what we think of as high T or low T doesn't have to be in reference to absolute levels. It's just as useful and valid to refer to a T level that's high "for males" or "for females," or that's high relative to what was seen in that individual a minute, hour, month, or three years ago. Consider, for example, the recent finding that one in six elite male athletes have testosterone levels below the normal reference range. Given that these men were sampled a few hours after taking part in a major national or international athletic competition, we would hardly want to predict that athletes with modest T levels (in some cases below the average for female elite athletes) have little in the way of competitive inclinations.[38]

Another major departure from the T-Rex view of testosterone comes from a well-established principle in behavioural endocrinology: hormones don't *cause* behaviour, but rather only make a particular response more likely. As Adkins-Regan explains:

> Hormones are one of several factors that go into the nervous system's decision. They may change the thresholds for other factors that enter into the decision (for example, thresholds for responding to stimuli from another animal) but are not normally the sole triggering agent.[39]

That is, rather than being a king who issues orders, T is just another voice in a group decision-making process. This, when you think about it, is extremely sensible. Even for animals in which the social situations encountered may seem trivially simple compared with the soap operas of human existence, there are still subtleties of context to be considered. The animal whose philosophy was to take a strictly hormone-response approach to the world would soon find itself in trouble. How an animal *actually* responds to a particular stimulus, like a potential mate or intruder, isn't determined by its hormonal state, but depends on social context: What is the relative status of everyone involved, who else is around, where is the encounter taking place?[40]

In fact, we already saw a vivid example of this in the cichlid castration study: the dominant social status of the castrated territorial fish trumped the higher T levels of his contender. Another demonstration of the same principle comes from a study of talapoin monkeys. Here, the captive community under investigation included both intact and castrated males, and the latter were periodically treated with whopping doses of testosterone to see the effects on social dominance (measured by aggression towards others in the group). Although this T dosing did increase the aggressive behaviour of the castrated males, this was invariably directed towards lower-ranking males. In other words, the relative social status of monkeys was a primary and powerful filter for whether T had any effect on aggressive behaviour. As a result, despite treatment that raised T levels above and beyond normal, "no animal rose in rank after hormone therapy."[41] In fact, there was no obvious relation between T and rank, with females usually, and castrated males often, ranking above intact males. T doesn't inevitably create a Rex.

An even more striking contrast with the T-Rex view comes from evidence that not only is T not sufficient to trigger a hormone-linked behaviour, but in some species it may not even be necessary. Take sexual behaviour. In many species, the hormonal coordination between fertility and mating is so tight that sex isn't even possible without adequate hormone production in the testes and ovaries.[42] Many male rodents, for instance, can't produce an adequate erection without intact testosterone-producing gonads, while in females, ovarian hormones control various bodily changes that make sex physically possible (like the seductive arched-back "lordosis" pose that makes female rats' otherwise inaccessible vaginas available for penetration). But in most primates, there are no such hormonal conditions. Hormones are instead linked with sexual motivation, rather than ability to copulate. According to Emory University behavioural neuroendocrinologist Kim Wallen:

> This separation of the ability to mate from sexual motivation allows social experience and social context to powerfully influence the expression of sexual behavior in nonhuman primates, both developmentally and in adulthood.[43]

In an elegant demonstration of this, Wallen looked at how a testosterone-suppressing treatment affected the sexual behaviour of male rhesus monkeys housed with females. In line with a link between T and competition, the treatment had a more severe effect on the sexual behaviour of monkeys grouped with several other males, and that therefore presumably had to compete for mating opportunities, compared with monkeys that enjoyed solo male status. (Interestingly, and a useful reminder that it's not just males that compete, something similar is seen in female rhesus monkeys, that are more likely to mate outside the fertile phase of their cycle if there's no female competition around.)[44] But even in the competitive multi-male situation, T suppression didn't always decrease sexual behaviour. Both prior sexual experience and higher rank served to protect against the effect of testosterone suppression. So although sexual activity ceased within the week in the lowest-ranking male, the sexual behaviour of a high-ranking sexually experienced male "was not detectably affected by testicular suppression."[45] This was despite his testosterone levels being at castration levels for eight weeks. Status and past experience overrode hormonal lack.

As a final indignity for the T-Rex account, not only is T neither a sufficient, nor necessary, cause of hormone-linked behaviour—sometimes it's not even really a cause at all. Recall the purpose of hormones: to "adjust behavior to circumstances and contexts."[46] To this end, T turns out to play "a key role" in helping animals tune their social behaviour to whatever social scene they find themselves in.[47] Although we're used to thinking of certain kinds of behaviour as "testosterone fuelled," in many cases it would make more sense to instead think of actions and situations as being "testosterone fuelling." Social context modulates T levels (up or down), which influences behaviour (presumably via changes in perception, motivation, and cognition), which influences social outcome, which influences T levels . . . and so on.[48]

Again, the cichlids offer a useful illustration. On first encounter, you'll remember, it seemed obvious that dominant fish were dominant *because* they had high androgen levels. But careful experimentation revealed that, in fact, dominant fish had high androgens because the stars of fate had aligned to make them dominant. When male cichlid fish are first put together, their androgen levels tell you nothing about who will end up high versus low in social status. Although from a T-Rex perspective we would assume that the fish with more androgens would "naturally" be more successful in clambering up the social ladder, this simply isn't the case: the relations between hormone and dominance go the other way around.[49] Only once the fish have had time to interact and jostle do correlations emerge, with successful fish producing more androgens. As Lisbon University behavioural neuroendocrinologist Rui Oliveira, lead author in this study, explains:

> Social information is translated into changes in levels of steroid hormones that in turn will modulate the neural network of behavior so that behavioral output is tuned according to the perceived social environment.[50]

In fact, the effects of the social world have even been seen at the genetic level, with social interactions changing androgen and oestrogen receptor expression in the brain.[51] Testosterone, in other words, is demoted from Rex to being a mere middle man that mediates the influence of the social world on the brain. Change the world, and you can change T—and the brain.

And, importantly, even in nonhuman animals it's subjective perception rather than physical reality that counts. Consider again the cichlids; in particular, one unfortunate male hailing from the Oliveira study. All but one of the socially dominant males in this study successfully established territories, and started to churn out androgens in greater abundance. But that one fish, despite winning about 70 per cent of his fights, failed to establish a territory. Intriguingly, the androgen production of this winning fish was an outlier on the graph, being down at the level of the group of vanquished fish. As Oliveira notes, "This suggests that it is the individual's perception of its status, rather than an objective measure of its dominance behavior, that triggers [androgen] production."[52]

Or consider a study of male marmoset monkeys, a monogamous species in which fathers are actively involved in parenting. Researchers measured T response to the ovulatory odours of unfamiliar females, and found that it depends on the male's family status. Single males showed testosterone elevations (as well as penile ones) in response to the sexually enticing smell. But to "family" males (those pair-bonded with offspring), this same stimulus apparently had little effect—perhaps because it represented a distraction rather than an opportunity—and their T levels remained unresponsive.[53]

In short, the picture painted so far is a far cry from the simple T-Rex view, in which testosterone fuels male competition in direct proportion to the stable quantity of it roiling in the bloodstream. We've already seen that competition is an important feature of females' lives too. And with the circulating level of T being just one variable in a complex system—and one in which the sexes may potentially use different means to reach similar ends—we can't assume that T is only important in males. T is also just one of many factors that feed into an animal's decision making. Social context and experience can override its influence on behaviour, or fill testosterone's small shoes in its absence. And finally, far from being a pure biological measure of hormonal sex, T *responds* to contexts and situations, meaning that whatever influence T has on the brain and behaviour can't be simply chalked up "to testosterone," a purely biological factor. T level or reactivity is inextricably intertwined with the individual's history and current subjective experience.

So what about ourselves?

In keeping with other animals, T likewise seems to help us to adapt our behaviour to "circumstances and contexts." So, when it comes to relatively enduring circumstances—basics like partnering and parenting—T levels seem to be in line with the principle of high T being linked with competition, and low T with nurturance. For instance, both women and men with interest in acquiring new sexual partners tend to have higher circulating T than do happily coupled (or happily single) counterparts, and the parents of young children have lower T than non-parents.[54] And while it's hard to disentangle cause and effect in this kind of realm—needless to say, scientists can't randomly allocate people to ten years of marriage, or a baby—this doesn't seem to be simply due to people with higher versus lower T levels being drawn to different lifestyles. For instance, the findings from a study of male air force veterans, who were brought into the lab for regular testing of hormone levels and to report on their marital status, "illustrate the dynamic nature of testosterone levels, elevated in the years surrounding divorce, and declining through the years surrounding marriage."[55] The authors speculate that this happens because:

> The marriage ceremony is the culmination of a more gradual period of courtship and engagement, in which a man accepts the support and consortship of his partner, removing himself from competition with other men for sexual partners.

> As a result . . . his testosterone declines. In contrast, impending divorce is a time of competition between spouses for children, for material possessions, and for self-respect. Also, it is a time when the divorcing husband may reenter the competitive arena for sexual partners.[56]

And an arrow of causality from caregiving to T-level change was clearly seen in a large-scale longitudinal study of fathers in the Philippines, led by University of Notre Dame biological anthropologist Lee Gettler. This study found that fatherhood reduced testosterone levels in men, and more so in fathers who spent more time physically caring for their infants.[57] Nor was this simply because lower-testosterone men were more likely to be nurturing fathers; rather, intimate caregiving itself lowered testosterone.

But also notice how we're *not* like other animals: our social constructions of gender bring a uniquely human dimension into the mix. As we've already seen, gender norms and patterns for sexual behaviour and parenting take on widely different forms across time and space. These cultural circumstances are surely entangled in women's and men's hormonal biology. Illustrating exactly this situation is a study that compared two neighbouring cultural groups in Tanzania—Hadza foragers and Datoga pastoralists—each with very different expectations of fathers. It found lower testosterone levels among fathers from the Hadza population in which paternal care was the cultural norm, compared with Datoga fathers among whom paternal care was typically minimal.[58]

By the way, lower T levels don't doom devoted husbands and fathers to a submissive or sexually sparkless life. Contrary to popular belief, in humans there's little convincing evidence for a significant link between baseline circulating T and social status, and most studies have failed to find relationships between T and sexual desire in healthy men with T levels within the normal range.[59] This may well be because competition and status are more intermittent and situational for us than for some other animals. (And as van Anders points out, sexual desire can also flow from feelings of love and intimacy.)[60] We don't, for instance, all simultaneously take two weeks annual leave, fight ferociously for the best homes in which to rear our children, then frenziedly mate. It would seem to make more sense, then, for T to obligingly rise or lower temporarily, as context demands, or as opportunity knocks.

But here again, social constructions of gender will shape both the situations people encounter, and their subjective meaning. We're used to thinking of testosterone as being a *cause* of gender, but what if the direction of that familiar pathway also needs to be reversed? Some ingenious recent research by van Anders and her colleagues has started to lay down the evidence.

Take a study in which van Anders and her team acquired one of those programmable crying, sleeping, gulping, pretend babies that high schools use to illustrate the vital fact that using contraception, however inconvenient it might seem at the time, is considerably less effortful than parenting.[61] One group of men were randomly allocated to a role we will imaginatively describe as "traditional man who lets the woman do the baby care." They were instructed to simply sit and listen to the baby cry. Another group of men, again randomly assigned, formed the experimental condition we will refer to as "traditional man who lets the woman do the baby care and is therefore woefully inexperienced in that demanding, acquired skill, but on this occasion has been left alone with a baby." These men were told to interact with the baby, but, cruelly, it was programmed to cry persistently, regardless of what they tried. The last group of men were what we will call "progressive dads": the baby was set to cry, but programmed to be consolable when trial and error led to the right kind of comfort. In this last group, testosterone levels dropped as their tender ministrations took their desired effect. But in the other two groups, faced with the challenge of a profoundly unhappy baby, particularly when they just sat and listened, their testosterone rose. In other words, ostensibly the same stimulus—a crying baby—affects T differently, depending on the person's ability to deal with the situation.[62] Now consider the fact that, outside the laboratory, the confidence and experience a person brings to the challenge of a crying baby is likely to be shaped by gendered expectations and experiences around child care. Claims that men don't have "the right hormones" for taking care of babies are cast in a whole new light.

A second study by the same research lab shifted the simulated context from home to work. This time, van Anders and her colleagues trained male and female actors to perform a workplace monologue in which they enacted power by firing an employee, measuring T levels both before and after.[63] Displaying power didn't significantly affect men's testosterone levels overall. However, it did significantly increase T in the women. The interesting implication that the researchers drew from these findings is that social constructions of gender, which make displays of power both more likely and acceptable for men, contribute to the female/male gap in circulating T levels. "Gendered behavior modulates testosterone," the researchers conclude, pointing to "an additional reason for differences in testosterone: the understudied role of nurture."[64] (That's right, guys. Let women take your high-power jobs, and before you know it they'll be taking your hormone too.)

In this study, the social context was unambiguous: an incontestable power monologue. But often situations are more subjective, and taking this into account may be helpful for making sense of the tangle of results from studies of relations between testosterone and competitive behaviour on the sports field and in the lab. In these contexts, participants are usually unsure of just how tough the competition is that they're up against, or how things will turn out. At first, following a few null findings with women, it was prematurely assumed that only men show testosterone reactivity to competitive situations.[65] But as more data piled in, what emerges is, as one review recently summarizes it:

> an inconsistent pattern in both sexes, with T levels increasing in winners and decreasing in losers . . . increasing both

in winners and losers, or not showing significant changes in response to the competitive event.[66]

A sceptical conclusion would be that testosterone isn't doing much at all here.[67] But maybe, the authors of the review suggest, these inconsistencies are created by the different lenses through which people perceive a competitive situation. As endocrinologists Gonçalo and Rui Oliveira put it:

> The same exact event may elicit different responses, depending on the way it is appraised by different individuals or by the same individual at different moments in time (e.g., in different social contexts).[68]

Factors that researchers suggest could be important for how and when T responds to competitive situations include your appraisal of the skills of your opponent, the explanations you come up with to explain why you won or lost, how familiar you are with the "where" and "who" of the competitive situation, and your underlying motivations.[69] This is where gender can make its entrance. We might, for instance, expect gender to influence T reactivity via the stereotypes that help create expectations or shape explanations for success or failure, the existing inequalities that create double standards for performance, and gendered experiences and social networks. After all, as we saw in the previous chapter, a different domain (masculine versus more neutral or feminine), a different cultural background (for example, patriarchal versus matrilineal), or even a different framing of the same competitive context (sports news assistant versus administrative assistant) can eliminate sex differences in willingness to compete.

In fact, research with men has already shown the effects of culture and social constructions of gender on hormonal biology. Take the enduring effects of a successful ten-year Fast Track intervention programme in the United States. Targeted at boys at high risk of later antisocial behaviour, it "was designed to build social competencies and self-regulatory skills that enable children to respond more calmly and less vociferously to provocation."[70] Some participants received the intensive, decade-long intervention; a matched control group didn't. Many years later, when the participants were in their mid-twenties, Nipissing University social neuroendocrinologist Justin Carré and colleagues invited about seventy men from the study into the lab, and tested the aggressiveness of their reaction to a provocation (supposedly, another participant vindictively stealing points from them in a game). The intervention group were less likely to retaliate against what they assumed was hostile behaviour on the part of a confederate, showing long-lasting effects of the intervention. But most interesting for our purposes is that they also showed less testosterone reactivity to the provocation, and this in part seemed to underlie their greater inclination to turn the other cheek in a competitive context. As Carré and colleagues conclude:

> Together, these results suggest that the Fast Track intervention creates persistent changes in psychological processes underpinning how individuals encode, interpret, and process social threat and provocation. These mental processes, in turn, influence the pattern of testosterone responses to provocation, which in turn influence aggressive behavior.[71]

A similar conclusion emerges from a classic experiment by University of Illinois psychologist Dov Cohen and colleagues.[72] This time the comparisons were between non-Hispanic white male students who hailed from either the northern or southern regions of the United States. In a series of experiments, groups of men were placed in a contrived challenge to their social status in which they were bumped in the shoulder by a male decoy. The decoy then added insult to injury by muttering an offensive word. To northern students, this event was of little consequence. But the southern students, raised in the lingering remnants of a culture strongly grounded in a man's honour and the importance of the respect given to him, tended to walk away feeling worried about the effect on their reputation as a man. It was a slighted southern group, too, who increased their aggressive and domineering behaviour afterward. And again, it was only disrespected southern men who showed testosterone increases in reaction to this small challenge to their status. The discussion section of the study reassures that the experimental manipulations didn't "produce any truly violent behavior."[73] But suppose an unfortunate confederate *had* been punched by an affronted Southerner. Would it make sense to blame the testosterone? Or to say that *boys will be boys*?

In *Testosterone*, Herbert concludes that "the human brain has had to devise multiple ways of regulating, channelling, and optimizing the powerful effects of testosterone on male behaviour through laws, religion, and customs."[74] But there is no "real," "original," or "intended" testosterone level or reactivity with which civilization then interferes. As Wade remarks:

> Hormones, then, are not part of a biological program that influences us to act out the desires of our ancestors. They are a dynamic part of our biology designed to give us the ability to respond to the physical, social, and cultural environment.[75]

The studies showcased in this chapter vividly illustrate Agustín Fuentes's observation (echoing many feminist scientists),[76] that "when we think about humans it is a mistake to think that our biology exists without our cultural experience and that our cultural selves are not constantly entangled with our biology."[77] And culture seems to enjoy the upper hand.

OVER THE PAST EIGHT YEARS or so, I've taken part in a lot of discussions about how to increase sex equality in the workplace. Here, I would like to clearly state for the record that castration has never been mentioned as a possible solution. (Not even in the Top Secret Feminist Meetings where we plot our global military coup.) For organizations looking to increase the representation of women at senior levels, HR would eliminate castration immediately on obvious ethical and legal grounds. However, it can also be dismissed for scientific reasons. There's nothing in the scientific literature to suggest that castration—even in conjunction with testosterone patches for women—would provide a powerful biological shortcut to equality. It doesn't work for fish or monkeys, so why would it work for us? The very factors—status, experience, meaning—that become entangled with, moderate, substitute for, and override testosterone are human specialties *par excellence*—and no king emerges out of these complex interrelations.

What *would* work, the research instead suggests, are major and sustained interventions on status, experience, and what a particular situation *means* to the individuals involved. This, it's worth pointing out, poses a much more difficult challenge than administering hormone boosters or blockers.[78] The "broad tapestry"[79] of gender is tightly woven, and thick with redundancy: you can loosen one thread, but the others will still hold everything in place. A cichlid fish has size and flashy colouring to mark status. We have stereotypes that stain every encounter, clothing, language, salaries, titles, awards, media, legislation, norms, stigma, jokes, art, religion . . . the list of phenomena that make up our rich, gendered cultures goes on and on.

That's a lot of social construction to reconstruct. The big mistake is to confuse the persistence of the status quo with the dictates of testosterone.

CHAPTER 7

THE MYTH OF THE LEHMAN SISTERS

There's a very simple reason why most financial traders are young(ish) men. The nature of trading incorporates all the features for which young males are biologically adapted. . . . The whole set-up seems to have been designed for young men. All the actions of testosterone are echoed by the qualities of a successful trader. It does seem remarkable that the artificial world of financial trading should so suit the innate characteristics of young males.

—JOE HERBERT, *Testosterone*[1]

"IF LEHMAN BROTHERS HAD BEEN LEHMAN SISTERS, RUN BY WOMEN instead of men, would the credit crunch have happened?"[2] This question, posed by a *Guardian* business editor, triggered a "frenzied engagement in the international media with the gender question in international finance."[3] Some commentators, drawing on research reporting links between testosterone levels and risk taking, argued an urgent need for a greater "diversity of hormones":[4] more women (and older men) would make for less testosterone. Article headlines and interviewees repeatedly invoke T-Rex, calling for more "mistresses of the universe"[5] to bring some much-needed financial conservatism to the "testosterone rules"[6] world of business.

By now, the argument will be tediously familiar. Men, thanks to past evolutionary pressures, take risks in order to acquire the resources and status that led to reproductive success in our ancestral past. But fast-forward those Stone Age "male brains" to twenty-first-century global finance and the "evolutionary hangover" creates havoc, as Nicholas Kristof summarizes the view in the *New York Times*.[7] Enter subprime mortgages and credit derivatives, and with the benefit of hindsight it became clear just how dangerous it was to have all that testosterone in charge with barely a woman in sight. (Fully clothed ones, that is.)

It may seem a nice tribute to women to suggest that the world's financial system might not have been brought to its knees if only more of their representatives had been around. We should certainly take a moment to appreciate the contrast with a *New York Times* article published about a century earlier, reporting a movement among broker firms to forbid women from frequenting their offices.[8] As one, apparently typical, letter sent out to female clients by a Broadway firm of stock brokers explained, this was because their more valued customers "consider it undignified for women to frequent brokers' offices." Indeed, a woman is "a nuisance anywhere outside of her own home," one broker pointed out. Women, it was noted, not only lacked the "business instinct"—males presumably bursting forth from the womb in a pin-striped birthday suit with an innate understanding of finance—but were also incapable of acquiring it. It's certainly a long way from this to the May 2010 *Time* magazine cover featuring the female financial regulators—Elizabeth Warren, Sheila Bair, and Mary Schapiro—"charged with cleaning up the mess"[9] made by *you know who*.

There is a drawback, however, to belonging to the sex biologically suited to play the immobilizer of the "Big Swinging"[10] organs of finance. A risk-averse temperament may be good for the world economy, but bad for one's own personal finances. You don't have to be an analyst of distributions of wealth to be aware that the exercise of liquidating one's assets into a heap of dollar bills then gloatingly hurling oneself onto it would be a more comfortable experience for men, on average, than for women. Annual "rich lists" unfailingly confirm that it is almost exclusively men who would be at highest risk of suffocation during such an exercise. Historically, this inequality was easy enough to explain. Beyond advising a female client to marry well, even the most talented financial adviser would struggle to assist the wealth building of someone excluded from higher education, legally unable to own property and securities, and restricted to only the most low-paid occupations. However, these external barriers were dismantled some time ago, and many researchers have started to look to internal factors—like prenatal and circulating testosterone—to explain what has been described as "fundamental differences" in risk preferences, as one much-cited review put it.[11]

But by this point in the book you may be justifiably sceptical that T can create a "fundamental" divergence in the financial decision-making styles of women and men. As we've already seen, a major challenge to a T-Rex view of sex differences is the typical extent of the overlap in the ways that men and women, on average, behave. Of all the qualities we possess, was it really on those relating to finance that sexual selection acted with the most vigour, way back in the Pleistocene?

IN A HELPFULLY FORENSIC analysis of the scientific literature, the economist Julie Nelson reviewed eighteen studies of sex differences in financial risk taking from (and representative of) the economics literature.[12] Some of these studies used the lottery tasks so beloved of economists, in which people make a series of choices between, say, a sure $5 or a 50 per cent chance of winning $10. Others asked people to report on their preferences for financial risk taking in real life, or looked at how they allocated their actual financial assets among more and less risky options (like stocks versus bonds). As you might already anticipate, the effect sizes for these differences were, with a few exceptions, generally quite modest, with several null results (that is, no sex differences), and even two findings of greater *female* financial risk taking.[13]

How do we get from this overlap between the sexes to claims of a fundamental difference? By way of partial explanation, researchers often summarize results from earlier studies in an inaccurately stereotype-consistent way, observes Nelson. Researchers also tend to emphasize their own findings that are consistent with the stereotype of male risk takers, while downplaying (even sometimes to the extent of more or less ignoring) results that aren't. This aroused Nelson's suspicion that researchers are "tending to 'find' results that confirm socially held prior beliefs"[14]—a classic case of "confirmation bias." If these results are more likely to be published, the scientific literature becomes skewed towards the expected conclusion.

As it happens, there's a way of presenting data, called the funnel plot, that indicates whether or not the scientific literature is biased in this way.[15] (If statistics don't excite you, feel free to skip straight to the probably unsurprising conclusion in the last sentence of this paragraph.) You plot the data points from all your studies according to the effect sizes, running along the horizontal axis, and the sample size (roughly)[16] running up the vertical axis. Why do this? The results from very large studies, being more "precise," should tend to cluster close to the "true" size of the effect. Smaller studies by contrast, being subject to more random error because of their small, idiosyncratic samples, will be scattered over a wider range of effect sizes. Some small studies will greatly overestimate a difference; others will greatly underestimate it (or even "flip" it in the wrong direction). The next part is simple but brilliant. If there *isn't* publication bias towards reports of greater male risk taking, these over- and underestimates of the sex difference should be symmetrical around the "true" value indicated by the very large studies. This, with quite a bit of imagination, will make the plot of the data look like an upside-down funnel. (Personally, my vote would have been to call it the candlestick plot, but I wasn't consulted.) But if there *is* bias, then there will be an empty area in the plot where the smaller samples that underestimated the difference, found no differences, or yielded greater *female* risk taking should be. In other words, the overestimates of male risk taking get published, but various kinds of "underestimates" do not. When Nelson plotted the data she'd been examining, this is exactly what she found: "Confirmation bias is strongly indicated."[17]

This bias makes it misleading to draw conclusions from the entire literature, which inflates the size of the sex difference overall. So what differences in size did Nelson see when she looked only at the more precise results from the eight largest studies?[18] These included a large-scale newspaper survey that asked tens of thousands of respondents a lottery question, and two large-scale analyses of actual investments (thousands of retirement portfolios, and more than thirty-five thousand stock investment accounts).[19] Nelson's best estimate based on these eight most precise studies was an effect size of about 0.13. This translates into about a 54 per cent probability of a man picked at random being more financially risk taking than a woman picked at random. When you consider the conclusions of Chapter 5 regarding the often-gendered factors that are so important in explaining differences in people's risk-taking behaviour—like knowledge, familiarity, past experience, and gender norms that associate risk taking with masculinity—it begins to seem almost surprising that the difference is so small. Consider, too, that a person's wealth, likely future wealth, and financial security understandably affect the kinds of financial risks they're likely to be willing to make.[20] Although researchers can and do take account of men's and women's current wealth and earnings when comparing their risk taking, in a society in which men are more likely to be promoted, and women are more likely to suffer the career costs of caring for children and elderly parents, the financial trajectories and expectations of even initially comparable women and men are unlikely to be the same.

Sex differences in financial risk taking aren't just small, they're also conditional, showing themselves in some tasks, samples, and contexts, but not in others. One study, for instance, found no differences when the typical abstract lottery was expressed in the real-world context of investment decisions.[21] Similarly, Warwick Business School's Ivo Vlaev and colleagues found no sex differences when lotteries were put into real-world contexts. In among the traditional abstract lotteries, they presented equivalent ones in terms of pension, salary, mortgage, and insurance contexts. For example, the student participants were asked to choose between, say, a job offering £30 daily payment for sure, versus a riskier but potentially more lucrative job offering a 50 per cent chance of earning £100, and 50 per cent chance of earning nothing. Overall, they found no differences between the sexes.[22]

Another complication for the claim that men and women are fundamentally different in their approach to finances is that, as with other domains of risk taking (as we saw in Chapter 5) "this sample of females" and "this sample of males" shouldn't be taken to stand in for *all* females and males. Evolutionary scientists Joseph Henrich and Richard McElreath used a traditional lottery task to compare financial risk taking in three groups far removed from the Western students typically favoured in these kinds of experiments.[23] These were communities of small-scale farmers in Chile and Tanzania (the Mapuche and the Sangu, respectively), as well as a nearby Chilean non-farming community (the Huinca). Although cultural background made a difference—on average, the Mapuche and Sangu were risk prone, while the Huinca were risk averse—in none of the groups was being male per se linked with risk-taking propensity. Likewise, sex didn't explain any variation in financial risk taking in a study of more than four hundred Chinese participants sampled from the general population.[24] Meanwhile, another cross-cultural study comparing the matrilineal and patriarchal societies of the Khasi in India and the Maasai in Tanzania, respectively, failed to find any sex differences at all in risk taking on standard lottery and investment games.[25]

Emerging as a potentially important factor that might help to explain cross-cultural differences are the gender relations in the society from which participants are drawn. Thus, Binglin Gong and Chun-Lei Yang compared risk taking on a lottery task in the matrilineal Mosuo (in which the head of the household is also traditionally female) and the patriarchal Yi.[26] Although in both societies the females bet less than the men, the gap was considerably smaller in the matrilineal Mosuo. Similarly, University of Los Andes economist Juan-Camilo Cárdenas and colleagues found that the gender gap in financial risk taking was smaller in Swedish children than in children from Columbia, a country that ranks much lower than Sweden on various macroeconomic indices of gender equality.[27] There's even some evidence that single-sex environments may encourage greater risk taking in British girls and young women.[28]

The problems continue when you turn from lotteries to other kinds of financial risk tasks.[29] By now, you may have the impression that economists approach the topic of risk taking as though they once overheard a prestigious colleague comment that "Life's a lottery," and took it literally. But as we all know, the majority of financial decisions do not closely resemble economists' lottery tasks. Warren Buffett did not make his billions contemplating whether to pursue Option A with a guaranteed return of $2, or Option B with a 30 per cent chance of a $4 return. Nor do bosses fix employees with a gimlet eye and, slamming a coin onto the table with a provocative cry of "Heads or tails, Professor Massoud?" invite them to take a 50/50 gamble on a $15 per annum raise, or stick with the guaranteed $5 rise on offer.

One obvious difference is that people typically don't know the precise odds of the different hands they can be dealt by fate. This is also the case for two risk tasks popular with psychologists. In one (the Balloon Analogue Risk Task), participants decide how many times to "inflate" a virtual balloon with a pump, with 5 cents earned for every successful pump. At an unknown point, however, the balloon will explode and all the money is lost. A meta-analysis found that males are modestly more risk taking than females on average in this task.[30] However, precisely the opposite is the case for the second popular risk-taking task, the Iowa Gambling Task. Individuals choose between decks of cards that are either high risk (high rewards but also higher losses) and less advantageous in the long run, or lower-risk (low rewards and losses).[31] Although over time most people shift to the lower-risk packs, women are a bit more likely than men to continue to try their luck with the high-risk packs.

Another obvious and particularly important difference between laboratory tasks and real-world financial risks is the sums of money at stake. In the Balloon Task, financial rewards are scraping around the dollar mark, while in the Iowa Gambling Task the "dollars" lost and won are purely hypothetical. Largesse also tends to be absent in studies run by economists: pay-offs are usually small, hypothetical, or restricted to one particular gamble chosen at random. It's therefore noteworthy that in one of the few lottery experiments that compared risk preferences for trivial versus substantial pay-offs, the sex differences seen in the typical "small-change" version of the task disappeared when nontrivial sums of money were at stake.[32] An interesting perspective on why this might be was offered by anthropologist Henrich and his colleagues. They suggest that

> when actual economic stakes are 0 (hypothetical), all kinds of other concerns come to predominate in the decision process. Informants may be concerned with what the ethnographer will think of them or what other people will infer about them from their decisions.

In their own cross-cultural research, Henrich and colleagues therefore use large stakes "to focus the informants' attention on the game payoffs rather than on exogenous social concerns."[33] As you'll recall, sex didn't predict financial risk taking when this research method was used with the Mapuche, Sangu, and Huinca communities of Chile and Tanzania. Their research protocol is also in perfect keeping with Cass Sunstein's argument (first met in Chapter 5) that the consequences of a decision for one's self-concept and reputation are vital ingredients in the recipe from which preferences emerge.

This aspect of the decision-making context is something that economists, in particular, have not been especially interested in. It was only at the turn of the twenty-first century, in a groundbreaking economics article written by Nobel Prize–winning economist George Akerlof and fellow economist Rachel Kranton, that the concept that social identity and norms have a motivating effect on behaviour was formally introduced to economists.[34] "What people care about, and how much they care about it, depends in part on their identity,"[35] they observe. These "identities and norms derive from the social setting. . . . [S]ocial context matters."[36]

To a social psychologist, this is an almost comically belated revelation: a *little* bit as if only recently a landmark social psychology article introduced colleagues to the concept of money, and its remarkable influence on people's preferences and behaviour. But better late than never, and while this area of research is still in a preliminary state, the identities and norms at play in a particular context do seem to influence financial risk taking. For instance, when negative stereotypes about female mathematical ability are made salient (this can be done both by drawing attention to the person's female identity, as well as by highlighting the "masculine" nature of the task), it can impair girls' and women's interest and performance in mathematics—a phenomenon known as "stereotype threat."[37] In one study, women were more risk averse than men when they were required to record their sex before participating in a gambling task that was described as a test of mathematical, logical, and rational reasoning abilities. However, when exactly the same task was instead described as "puzzle solving" (and participants didn't record their sex beforehand), women were just as risk taking as the men.[38]

While this study manipulated the relevance of gender identity by framing the task as one of cool, rational calculation, risk taking itself is of course a key stereotypical trait of masculinity. According to popular imagination, for instance, the successful entrepreneur doesn't just have the necessary skills, resources, and business connections; he is also a masculine hero who laughs boldly in the face of financial risk. Perhaps this is part of the reason pitches made by male entrepreneurs are evaluated more positively than those given by female entrepreneurs—even when the content is identical.[39] Women can also be put off by this portrait, Binghampton University academic Vishal Gupta and colleagues have found. For example, when Turkish MBA students were shown either a (fictional) general news piece about entrepreneurship, or one describing entrepreneurs in a stereotypically masculine way (for example, as *aggressive, risk taking*, and *autonomous*), the male students later evaluated a potential business opportunity more positively than the female students, on average. But when, in two further conditions of the same study, that business opportunity was preceded instead by either a gender-neutral (*creative, well-informed*) or feminine (*caring, making relationships*) description of entrepreneurs, women were as likely, or even more likely, respectively, to see a business opportunity in the complex business case they then analysed.[40] Gupta and colleagues have also found that, across three countries, both women and men who reported having more "masculine" traits showed greater entrepreneurial intentions.[41] Interestingly, two other studies similarly found that men and women who report being more "masculine" in personality also score higher on measures of financial risk taking. Unsurprisingly, men report a greater number of masculine traits as characteristic of themselves, and this explains some of the gap in risk taking between men and women.[42] But as both research groups point out, while being biologically male or female is fixed, how masculine men and women see themselves is not. In fact, in the United States, the "masculinity gap" has been closing over time, in step with women's changing roles and status in society.[43]

If risk taking is an integral part of a masculine identity, then we can predict that men should take greater financial risks when that identity, or the norms associated with it, are made salient. Viennese academics Katja Meier-Pesti and Elfriede Penz found exactly that. They primed young women and men with either masculine, feminine, or (in a control condition) gender-neutral stimuli. Men primed with masculinity gave the most risk-tolerant responses on a questionnaire assessing attitudes towards risk taking in investments.[44] A more recent study also explored the importance of masculine identity for financial risk taking, by exploiting a rather depressing phenomenon known as the "failure-as-an-asset" effect. It turns out that presenting men with evidence that they have done poorly at something at which women tend to excel provides a little boost to their self-esteem, because incompetence in low-status femininity helps establish high-status manliness. Remarkably, failure in feminine domains is also perceived as an asset by onlookers. Fictitious male job applicants who reveal weakness in a "feminine" domain (like dancing, or in a form of intelligence in which women supposedly excel) are seen as more masculine, and thus more likely to succeed in high-level roles, compared with men who "lack" incompetency in femininity.[45] Expanding on this phenomenon, University of Kassel psychologist Marc-André Reinhard and colleagues found that giving men failure-as-an-asset feedback increased their self-reported interest in risky activities, as well as the amount they were prepared to invest in a gamble. This shift seemed to be brought about by greater identification with being male.[46] Interestingly, the investments of men who were told either that they'd done poorly on a masculine test, or well on a feminine one, made investments that were no riskier than those of women.

In apparent contradiction, research conducted at the University of South Florida found that young men take greater financial risks after a *threat* to their masculinity. (The psychological castration was achieved by asking a group of the men to try a florally fragranced hand cream.)[47] However, the contrast with Reinhard and colleagues' findings may lie in the private versus public nature of the risk taking. Masculinity threats only had an effect on financial decisions made publicly, suggesting that costly displays of masculinity in response to a threat of manhood are only worth it if they serve a face-saving function.[48]

Although we have to be careful that findings like these are robust and replicable, they have an important implication, as Nelson points out:

> Differences that may appear at a cursory level to be due to "essential" differences between the sexes may in fact be due (in part or completely) to some additional, confounding variable, such as societal pressures to conform to gender expectations or locations in a social hierarchy of power, or may no longer be seen when the sampling universe is broadened.[49]

Yet researchers may nonetheless treat results as though they reflect categorical, Mars *versus* Venus differences, Nelson goes on to point out. For example, a four-country comparison of the risk preferences of female and male asset managers revealed only marginal, scattered, and unsystematic differences across the four countries. Even for the most sizable difference in risk preference (which was seen in Italy), the possibility of creating a perfect match by pairing a female client with a female fund manager was only 38 per cent, compared with a 25 per cent chance of a successful match if the customer employed a male manager instead. Nonetheless, the study authors suggest that "female fund managers may be better suited to female customers."[50] As Nelson wryly notes, given that economists assume that financial risk-taking preference can be readily assessed with a few simple questions, why not simply ask clients what they want? It's a bit like a restaurant manager learning that women are slightly more likely than men to order fish than steak, then telling waiters to use a customer's sex as a guide to what meal to place in front of them.

Why might researchers make this kind of conceptual slide from small average differences to fundamental difference? Is it because, implicitly, they join so many others in assuming that the sexes just *are* essentially different? As Nelson explains:

> The attribution of (on average) different psycho-social behaviors to (fundamental) sex differences in hormones and/or brain structure, further explained as caused by differences in evolutionary pressures on bodies with different reproductive roles, can currently be found in many studies.[51]

Sound familiar, at all? This assumption, in turn, makes it easier to neglect important features of the data: the vanishingly small size of sex differences in financial risk taking, and the dependence of those differences on who is being tested, the kind of task, and the social context. As we've already learned, these details matter a lot for the kind of explanations we reach for. If we say that "men are financial risk takers, and women are financially risk averse," then men's higher testosterone exposure looks like a plausible cause of that difference. But to echo questions asked earlier, how does the substantial sex difference in testosterone translate into such modest behavioural differences? How does T make men more risk taking when a gamble is framed in an abstract way, but not in a concrete salary context? Or when stakes are trivial, but not necessarily when they're substantial enough for a loss to sting? How do sex differences in testosterone make young North American men more risk taking than their female counterparts, but not the men of China, Mapuche, Sangu, or Huinca? How does T make men inflate more risky balloons, but select fewer risky cards?

These are questions to bear in mind as the research seeking to link T with financial risk taking proceeds apace. One growing line of investigation tries to find links between financial risk taking and a measure known as digit ratio. Digit ratio is the relative length of the second to the fourth finger and on average men's digit ratio is smaller than women's.[52] Digit ratio is often popular with researchers because it's so easy to measure, and supposedly reflects prenatal exposure to testosterone, although whether or not there's adequate evidence for this is contentious. (One set of researchers, for instance, describe digit ratio as "a putative, not yet sufficiently validated marker of prenatal testosterone.")[53] But even if it's a reasonable measure for comparing prenatal testosterone *between* groups, digit ratio "may be much less useful" as an index for individuals *within* a group, as Herbert explains.[54] (It's just too "noisy" a measure: a bit like using a person's height as a proxy for their early nutrition, on the grounds that people who were fed well as kids are taller, on average, than those who were underfed.)

But setting all this aside, from a Testosterone Rex perspective, it's not hard to understand why researchers might be interested in looking for a correlation between digit ratio and financial risk taking. The traditional view of sex differences in the brain (as we saw in Chapter 4) is that the high levels of T produced by the newly developed testes of unborn boys plays a singularly important role in creating discreet "masculinized" circuits in the brain. These circuits, especially when activated by the higher levels of T at pubescence and beyond, are the basis of distinctly male sexually selected mating behaviour: like fighting off other contenders for the cosiest cave; hunting ferocious, meaty prey; and today, apparently, buying high-risk biotech stocks. Put these outdated assumptions together with another, that we met in Chapter 5—that risk taking is a stable, masculine personality trait—and the long chain of reasoning is complete. A person with a lower, more male-typical digit ratio will have a more "male brain"; a person with a more "male brain" will be more masculine; a person who is more masculine will be more risk taking; and a person who is more risk taking will be more financially risk taking. Thus, someone who, according to their digit ratio, was putatively exposed to more prenatal testosterone will, many decades later, be reliably more likely to say: "What the hell, I'll take the 30 per cent odds of winning one dollar, rather than the twenty cents for sure."

You'll already be aware from earlier chapters of the weakness in several links in this chain. The assumption that resources and status—and therefore risk taking and competition—are distinctly *male* concerns in the struggle for reproductive success (and should therefore be wired into a "male brain") was dissected and found wanting in the first part of the book. In line with this abandonment of dichotomous "competitive males" versus "coy females" thinking in evolutionary biology is the shift in neuroscientific understanding of sex and the brain. The notion of discreet, T-induced "male" circuits is being replaced with a more complicated, interactive mishmash of factors, out of which emerge a variety of shifting "mosaics" of brain characteristics. This, in turn, fits nicely with what we know about sex differences in behaviour. These certainly exist but, again, in ways that create mosaics rather than categories. Put this all together, and it probably shouldn't surprise us too much that recent meta-analyses and a large-scale study failed to find convincing evidence for correlations between digit ratio and other supposedly quintessentially masculine behaviours: aggression,[55] sensation seeking,[56] dominance, and both aggressive and non-aggressive risk taking in adolescents.[57]

Along similar lines, as we saw in Chapter 5, although you might assume that your friend, Ankush—who goes skydiving every weekend—must therefore be "a risk taker" with a "male brain" (thanks to an abundance of testosterone prenatally and/or in adulthood), you may well later discover that Ankush's investments are all in government bonds. Looking for links between financial risk taking and testosterone exposure makes sense from the old perspective we met in Chapter 5, in which risk taking is assumed to be a stable, domain-general personality trait. But less so, however, from a more nuanced understanding of risk-taking behaviour as custom-made for each situation out of the "unruly amalgam" of factors, including social identity, norms, knowledge, past experience, social context, and the perceived risks and benefits in that particular domain. Which kind of risk taker do you expect to have the most male-typical digit ratio? And in which circumstances?

So how is the quest to find a robust, reliable link between digit ratio and financial risk taking progressing? A recent review politely describes the results as "equivocal."[58] There *are* positive findings here and there; but as the review authors explain, because of the number of different ways there are to look for links between digit ratio and behaviour, researchers have several "shots" at finding a statistically significant result. For example, researchers can use the measurements for the left or right hand, or an average of the two, and results can be looked at for the sexes separately or together. These various options, with just one measure of risk taking, yield no fewer than nine possible correlations to put to the test.

What about studies looking for correlations between circulating testosterone levels (in the blood or saliva) and financial risk taking? "Equivocal" is probably a pretty good one-word summary here too. For instance, depending on which study you look at, higher risk taking in lottery tasks is associated with higher T (in men; women weren't tested), with high or *low* T in both women and men, with higher T in men only (but only for risks taken to win money, not to avoid losing it), or with higher T in women, and men in the lower ranges of testosterone, or doesn't correlate with T at all in either women or men.[59] People with higher T *do* play more risky cards in the Iowa Gambling Task, and men with higher T (women weren't tested) play a riskier game in the Balloon task, but only if their cortisol (a stress hormone) level is low.[60] Meanwhile, a recent study that used a trading simulation found no relations between testosterone levels and risky trading behaviour in either sex.[61] As for real-world financial risk taking, one study found that male MBA students with considerable experience in the risky business endeavour of new venture creation had significantly higher testosterone levels than other students (there weren't enough women to include in the analyses).[62] But another study of MBA students found only a very small positive correlation between circulating testosterone levels and the choice of a risky career in finance, which disappeared when the researchers took account of the sex of the participants.[63] (If men both have higher levels of testosterone and, for possibly completely unrelated reasons, are more likely to be interested in a career in finance, then you will find a correlation between testosterone and the choice of a finance career even if the two are actually unrelated within either sex.)

How do you get from this weak mishmash of results to the idea that there's "too much testosterone"[64] on Wall Street? It probably helps that this is a story that fits perfectly with the T-Rex view of sex differences. But media reports also often refer to the research of Cambridge University's John Coates, a former trader turned neuroscientist, and Joe Herbert, whose work links higher circulating testosterone levels in male traders to higher profits on the trading floor.[65] At first glance, Coates's and Herbert's findings seem to indicate a need for *more* testosterone on Wall Street, not less, since men with more T do better. But Coates argues that the effects of testosterone can become detrimental in certain contexts. In a bull market in which share prices are rising, traders' testosterone levels get higher and higher as they make more and more money (known as the "winner effect" in animal research that finds testosterone increases in animals following a win in a competitive interaction). At a certain point though, "testosterone shifts traders' risk profiles to become overly aggressive."[66]

In line with Coates's account, a recent study did indeed find that men's testosterone levels rise after winning a game, and that this increase in testosterone is positively correlated with greater financial risk taking.[67] (Women weren't tested.) But this point—that a person's experiences influence their testosterone level—is critical to bear in mind when thinking about the results with the traders. As we saw in the previous chapter, T isn't a pure biological measure, but is entangled with the individual's history and current social context. This makes it impossible to say from the trading floor study that higher T levels directly cause greater financial risk taking. By way of a mundane alternative explanation, young men's testosterone levels are reduced by interrupted sleep,[68] and a poor night's sleep could plausibly interfere with the complex and time-pressured financial decision making of the trading floor. Or perhaps on certain days the traders learned useful information in their morning briefing that both boosted testosterone and increased their chances of successful trading. To show that higher T levels *cause* greater financial risk taking, you need to manipulate people's T levels, then look at the effects on behaviour. To date, only a handful of studies have done this. So far the picture is pretty mixed and mostly negative.[69] However, the recent study that used the trading simulation to measure risk taking found that although T levels were unrelated to financial risk taking (in either men or women), testosterone administration did increase men's investment in high variance (that is, more risky) stocks. (Women weren't included in this part of the study.)[70] What we seem to be left with, then, is little evidence that absolute testosterone level per se is related to financial risk taking, but the tentative possibility that it's *change* in T that's important.

If so, then what is the relevance of men's higher absolute levels? Unfortunately for the "Lehman Sisters hypothesis," it's impossible to draw any conclusions about women, testosterone, and trading tendencies from data collected solely from men. Coates realizes this, of course, but suggests that because of their lower T levels, female traders don't show the same hormonal reactivity to market activity: for instance, he argues that they are less susceptible to the "winner effect."[71] Yet this seems to be simply speculation, perhaps inspired by a "rutting stag" model of sexual selection that, as we saw in the first part of the book, applies poorly to humans. As we saw in the previous chapter, women's T levels *are* also sometimes responsive to competition, that level is just one part of a complex system, and in both sexes, T reactivity is inconsistent and conditioned by history, context, and norms.

The myth of the low-testosterone Lehman Sisters relegates women to the "mothering" roles of curtailing the excessive risk taking of male colleagues, and mopping up organizations' messes (a well-documented bias dubbed the "glass cliff" effect by Michelle Ryan).[72] As three leading business school academics point out in a letter to the *Financial Times*, while being "the first to argue for greater inclusiveness and a more diverse leadership to lead us out of this mess," claims that women are inherently more risk averse

> have little or no empirical support in a business context. These speculations also come with dangerous implications. Are men therefore better suited to managing growth or leading businesses through healthier economic times?[73]

That certainly seems to be the conclusion drawn by some. When asked by a journalist to "imagine what a world might be like maybe with no testosterone or if everyone had the same kind of levels as women?" Herbert replied, "Testosterone has got a bad press, but actually, it's responsible for a huge amount of get-up and go, of innovation, of drive, of motivation, of excitement." But only in men, apparently. "The suspicion," says Herbert, is that testosterone, "doesn't necessarily have the same effect" in women. After all, they "have a female brain, whereas a male brain is substantially different."[74]

WE'RE UNLIKELY TO FIND OUT any time soon how a "Lehman Sisters," or even a "Lehman Sister and Brother,"[75] would operate. One scholar describes the financial sector as "one of the few bastions of virtually uncontested masculine privilege remaining in the aftermath of feminism."[76] More equal representation of women at higher levels of the finance industry most likely *would* be beneficial. Lack of diversity is usually an alarm bell that people are being drawn from a limited talent pool that flatteringly reflects the image of those in charge. The "white male" effect described in Chapter 5 also provides a good object lesson in the importance of diverse backgrounds and identities for robust risk assessment. And as Nelson suggests, in a speculation that evokes the dismal "failure-as-an-asset" effect, greater senior female representation might go hand in hand with a much needed destigmatizing of positive "feminine" qualities:

> *Were* Wall Street firms and regulatory agencies such that they welcomed women and men as equal participants, this might indicate that societal gender stereotypes were breaking down. It might also be likely, then, that certain valuable characteristics and behaviours commonly stereotyped as feminine (such as carefulness) would be encouraged industry-wide, and inappropriate male-locker-room and cowboy-type behaviours frowned upon, to the benefit of the industry and society.[77]

However, there's currently little compelling evidence to suggest that this would be because women make financial decisions in a fundamentally different way from men, or because they would lower the average level of testosterone in those shiny, expensive buildings.

Last time I looked, it was largely taxpayers and society that, via "financial socialism,"[78] covered the costs of the decisions that brought about the global financial crisis. And, to my knowledge, there are currently no data investigating links between sex differences in testosterone and the taking of "risks" where benefits are reaped for oneself, but losses are underwritten by others.

PART THREE

FUTURE

It's such a chauvinistic sport. I know some of the owners were keen to kick me off Prince, and John Richards and Darren stuck strongly with me . . . I just can't say how grateful I am to them. I just want to say to everyone else to get stuffed because they think women aren't strong enough but we just beat the world.

—MICHELLE PAYNE, first female jockey to win the Melbourne Cup[1]

CHAPTER 8

VALE REX

Deeds Not Words

—Motto adopted by Emmeline Pankhurst[1]

A LITTLE WHILE AGO, BUYING FLOWERS AT A LOCAL SCHOOL MARKET, I overheard a conversation taking place at a nearby stall. The woman there was selling plastic knives for kids that, according to the marketing material on display, were guaranteed to keep little fingers 100 per cent safe. Having secured a two-knife deal with a family, the booth seller asked the daughter if she'd like a pink knife, and then asked her brother if he'd like a red knife or a blue one. "I'd like a pink one too," he said. As I enjoyed the moment, surprisingly, my eldest son ambled into the scene.

"If I manage to cut off a finger with one of your knives, can I have it for free?" he asked the booth seller. In reply, the woman irritably told him to leave her alone as she had work to do. *Yes, indeed,* I thought. *A busy schedule buttressing the gender divide with your pointless plastic crap.*

Anyone who has bought children's toys in the last few decades will not be surprised to learn that it is deemed necessary by some for children's knives to come colour-coded for sex. So are many toys, as apparently there are two kinds of children. Sometimes, the kind of child a toy is for is bluntly stated: particular aisles or product Web pages are explicitly designated as *for boys* or *for girls*. Other times, there are hints that are no less readable. A toy in bold, dark colours, featuring exclusively male figurines, packaged showing only boys having immense fun with it, surrounded by a wall of similarly masculine-coded products geared towards action, competition, dominance, and construction does not send the inclusive message that this is a toy for anyone, regardless of genitalia. Likewise, the notorious "pink aisle" is not the brainchild of marketing minds at pains to ensure that no child gets the sense that this toy isn't intended for the likes of him.[2]

Unsurprisingly, sex-segmented toy marketing has incited plenty of campaigns, and harsh criticism from parents, politicians, scientists, marketing professionals, and even children themselves.[3] But some dismiss this as misguided political correctness. For instance, in an *Atlantic* commentary sparked by a toy catalogue with photos of children playing in both traditional and counter-stereotypical ways (like a boy playing with a baby doll), Christina Hoff Sommers writes that "[Boys and girls] are different, and nothing short of radical and sustained behavior modification could significantly change their elemental play preferences."[4] Speaking from a marketer's perspective, Tom Knox, as chairman of DLKW Lowe, argues that "expecting marketers to ignore basic and profound differences in their audience seems ill-conceived and impractical." (Unconventionally, what's meant by "audience" here is presumably "people we hope will buy our stuff.") Knox suggests that "there will always be a place for gender-specific toys, gender-specifically marketed, in a way that celebrates gender diversity without undermining equality."[5] Similarly, in the same article, Helenor Gilmour, then head of consumer insight and brand development at DC Thomson, argues that "by failing to acknowledge these differences as marketers we would fail to understand our audiences effectively and deliver the services and products they want."

Some academics, meanwhile, bring an evolutionary flavour into the mix, suggesting that marketers are working from an instinctive grasp of our evolutionarily honed differences. In an article titled "Intuitive Evolutionary Perspectives in Marketing Practices," for instance, the authors observe that "some people may want little boys to be less competitive," but then rhetorically ask:

> But who is going to have more success in the marketplace, firms that appeal to young males' propensity to behave competitively with one another or those that appeal to males as nurturers . . . ?[6]

Likewise, in his book *The Evolutionary Bases of Consumption*, Concordia University Evolutionary Psychologist Gad Saad argues that "given their desire to maximize profits, [toy companies] develop products that are successful in exactly the same sex-specific manner across innumerable cultures."[7] This sentiment is echoed in the *Sunday Express* by journalist James Delingpole, who writes that "a toy business's job is to make profit not engage in social engineering." Some thoughtful readers might wonder why the *laissez-faire* philosophy of gender-neutral marketing is "social engineering," while toy aisles that dictate which toys are for whom are considered to be leaving things to take their natural course. But Delingpole has a further complaint. Gender-neutral marketing is futile, he says, because "those XX and XY chromosomes will out in the end."[8] In short, calls for gender-neutral toy marketing are seen by some as tantamount to demands that toy companies put themselves out of business by disrespecting boys' and girls' true natures.

A few years ago, in the frenzied lead-up to Christmas, the Australian Greens senator Larissa Waters catapulted herself into the heart of this debate by endorsing a campaign against gendered toy marketing.[9] Waters went further than the usual complaint that "no child's imagination should be limited by old-fashioned stereotypes." These "outdated stereotypes," she argued, "perpetuate gender inequality, which feeds into very serious problems such as domestic violence and the gender pay gap."[10]

The reaction was a timely reminder that to refer to gender debates as "spirited" can be like describing the surface of the sun as "warm." Waters was disparaged from the front page of the news to highest political office. The Australian's *Daily Telegraph's* cover headline announced a "Greens war on Barbie," in which the subheading's claim of evidence of political party insanity—"Now they're really off their dolly claiming kids' toys lead to domestic violence"—was accompanied by an image of Waters and a male Greens MP photoshopped onto the bodies of Barbie and GI Joe.[11] Well-known Australian child psychologist Michael Carr-Gregg commented that "these gender differences are hard wired," adding that "to argue that toys in any way relate to domestic violence is, I think, too far a stretch. It's a nail in the coffin of common sense."[12] One liberal senator suggested that Waters must have "consumed too much Christmas eggnog to come up with an idea like this."[13] And judging from the commentary on talk radio, it seemed that the prime minister of Australia at the time, Tony Abbott, spoke for many when he said that he didn't believe in "that kind of political correctness." His advice: "Let boys be boys, let girls be girls—that's always been my philosophy."[14]

The phrases used to defend gendered toy marketing are telling: "elemental play preferences"; "basic and profound differences"; "hard wired"; "those XX and XY chromosomes"; "sex-specific"; "celebrates gender diversity"; "let boys be boys, let girls be girls." The assumption is that boys are naturally, universally, and immutably drawn to "boy toys" because it is their evolved, timeless, biologically rooted nature to be risk taking, competitive, dominant, and to master the world. For the same reasons, girls are inexorably drawn to "girl toys," because it is in *their* nature to nurture others and to want to look attractive. So what is the problem with marketing that simply reflects and responds to those different natures, and what on Earth is the point of politically correct marketing that ignores them? What next? Ads trying to sell hockey sticks to cats?

From the Testosterone Rex perspective—sex as a powerful, potent, polarizing developmental force—this view makes perfect sense. But as we've seen, in the evolution of the science of sex and society, Testosterone Rex has not survived. As we saw in the first part of the book, both across and within species, biological sex doesn't have straightforward consequences for male and female roles. Sperm provisioning turns out not to be as biologically cheap as people still sometimes assume, nor competition and social dominance as irrelevant to females. Bateman's principles aren't obsolete, but nor are they omnipotent and omnipresent. Many different social, physiological, and ecological factors enter the mix, making sex roles dynamic, and even reversible.

This is especially clear when it comes to our own species: in the course of our evolutionary history, we've conspicuously failed to reach a species-wide consensus on "the" way to pair and raise children. Of course, every evolutionary account of humans acknowledges the major influence of the physical, social, and cultural environment on sexuality. But perhaps less recognized is that it has somehow come to pass that our sexual behaviour is uniquely uneconomical—we enjoy an unparalleled amount of non-reproductive sex. If humanity were a factory for producing babies, everyone would be fired. The considerable time and energy costs of our often unproductively nonreproductive sex points to its primary purpose no longer being reproduction, as we saw in Chapter 3. Understanding sexuality therefore requires us "to reconnect the genitals to the person," as Carol Tavris puts it.[15] For us, sex is not the means by which two well-matched reproductive potentials get together: we desire sexual activity *as a person*, in all our own unique, culturally crafted individuality, *with a person*, within our own particular cultural, social, and economic context. Presumably, that's why other cultures', and even acquaintances', pairings and preferences can be so mysterious.

A second important consequence of our inefficient sexual practices is the disruption to the supposedly universal principle that males' freedom from the labours of gestation, birth, and lactation should nudge them hard towards Maserati-driving, lady-magnetizing, baby-abandoning traits. Supposedly, it's the economics of reproduction that drives men, more than women, to succeed and sleep around, but it's easy to get too carried away in estimates of men's likely return on investment. In reality, in the absence of ecological, social, economic, and legal conditions that allow for harems, a man has to put in a hell of a lot of legwork to out-reproduce the steadfast husband and father. So why should we expect the reproductive potential of a tiny subset of men in a few pockets of history to be the foundation of a male essence—for there to be an incipient Genghis Khan in the sexuality and strivings of every male?

This diversity of possibilities for men illustrates the uniquely tricky developmental problem we humans have had to solve: "A newborn human must be ready to join any cultural group on Earth, and without knowing which," as evolutionary biologist Mark Pagel puts it.[16] And our genes don't know in advance what that cultural group's consensus will be on the appropriate roles for men and women. A baby girl could potentially be born into a society that expects her to play piano and embroider, study at a university, walk dozens of miles a day to fetch water, plant crops, tend animals, prepare animal skins, or hunt animals—and to grow up to live a life of chaste wedded monogamy, or to have two or three husbands simultaneously. For a baby boy, his destiny might involve crafting musical instruments, butchery, making nets, milking, pottery, investment banking, or intensive child care—and his future wife could be a thirteen-year-old girl or a thirty-year-old professional. Some kinds of future roles are more likely than others across societies, certainly, but all are possibilities.[17] And, regardless of our biological sex, life will likely demand we all, at some point, cherish and care for others; take risks; and compete for status, resources, and lovers.

Why, then, should we expect sexual selection to have fixed in our genes the recipe for a "female brain" and a "male brain" that creates distinct female natures and male natures, respectively? Certainly, the various genetic and hormonal facets of biological sex have to coordinate in ways that are (mostly) reliably directive when it comes to the reproductive system. But beyond the genitals, it would be useful for sex to be somewhat noncommittal, to play it by ear in its effects on brain and behaviour, to be pliant to the many other developmental resources it takes to build a person.

In other words, the developmental puzzle is *not* the one that Testosterone Rex so compellingly solves for us—how sex creates males who, beneath the cultural veneer, are timelessly, universally, and immutably like *this*; females like *that*. The *real* problem is how sex (usually) creates essentially different reproductive systems, while allowing the differences in men's and women's behaviour to be *non*-essential: overlapping and mosaic, instead of categorically different; conditional on context, not fixed; diverse, rather than uniform.

Some of the progress in working out how we achieve this neat trick comes from a major scientific shift, as we saw in the second part of the book. It has always seemed natural to ask: "How does this sex difference in the brain or hormones make females and males behave, think, or act differently?" That's the *only* question to ask when you're solving the red herring problem that Testosterone Rex explains. But a no less important question is how males and females can so often behave similarly, *despite* biological differences. When we notice that girls and women sometimes take risks and compete to the same degree as boys and men, when we realize that people have idiosyncratic mixes of "masculine" and "feminine" brain characteristics and gendered qualities, it becomes clear that biological sex can't have nearly as potent an effect on male and female behaviour as it does on male and female anatomy. And when we no longer assume that sex differences add up, and up, and up, we start to ask whether some sex differences are compensating for others, to make the sexes *similar*, not different.

A second scientific change is also helping to explain how sex can be such a helpfully light-handed and flexible influence in human development: a growing interest in how gender affects sex-linked factors, like testosterone. As Anne Fausto-Sterling advises, "think developmentally. Remember that living bodies are dynamic systems that develop and change in response to their social and historical contexts."[18] Testosterone changes bodies as well as brains, for instance, meaning that even when you measure a person's digit ratio, you don't just capture the effects of "sex," but potentially the cumulative effect of that person's more (or less) masculine appearance being responded to by others through a gendered lens. Nor do circulating T levels reflect pure sex. As we saw in Chapter 6, social context, experience, and subjective meaning can alter T levels—as well as override testosterone's influence on behaviour, or compensate for its absence. These often-gendered phenomena are a human specialty that, when we have the will to do so, we have a uniquely powerful capacity to change.

These gender constructions are a core part of our developmental system, bringing us to the final key to understanding the complex interrelations among sex, gender, and society. As we saw in Chapter 4, in animals, the developmental system—that legacy of place, parents, peers, and so on that every individual reliably inherits along with his or her genes—plays a crucial role in the development of adaptive behaviours.[19] In this regard, we are both like, and unlike, other animals. Our "complex and varied culture . . . resembles animal cultural traditions about as much as a Bach cantata resembles a gorilla beating on its chest," as Pagel observes.[20] Some evolutionary scientists argue that this uniquely human feature of our own developmental system is what makes our dazzling diversity of ways of life possible, in concert with another special, key human characteristic. This is an adaptation to learn from others in our social group. From the tender age of just two, we conform to the behaviour of our peers—notably, even other great apes don't "ape" each other in this way.[21] In particular, we're geared towards learning from those who are prestigious, successful, or similar to us in some important regard, with whom we come to identify, and from whom we learn, internalize, and gain our understanding of cultural norms.[22] Gender constructions penetrate just about every aspect of this cultural legacy. They aren't some dubious concept made up by gender scholars who don't believe in biology and evolution: they are *part* of both. Every newborn human inherits gender constructions as an obligatory part of their developmental system: gender stereotypes, ideology, roles, norms, and hierarchy are passed on via parents, peers, teachers, clothing, language, media, role models, organizations, schools, institutions, social inequalities . . . and, of course, toys.[23]

The T-Rex view of "boy toys" and "girl toys" is familiar from earlier in the chapter: the pink and blue categories reflect the preferences of "female brains" and "male brains" made distinctively different, in large part by the hand of testosterone. By way of evidence for this view, defenders of gendered toy marketing often refer to the more masculine preferences of girls with congenital adrenal hyperplasia (CAH). (As you might recall from Chapter 4, CAH is a condition in which very high levels of androgens are produced *in utero*.) And from here, it's just a few short steps to the conclusion that sex inequality is natural and inevitable. But since Testosterone Rex is extinct, we need another explanation of what's going on.

In the first year of life, baby boys and girls provide little in the way of evidence that their brains are tuned to different radio stations of life. For example, at birth, girls and boys are pretty similar overall in how interesting they find a face versus a mobile. Although a Cambridge University study found a statistically significant difference between the sexes,[24] even if you overlook important flaws in the method of this much-publicized study,[25] the differences are underwhelming. (Boys looked at the face 46 per cent of the time; girls, 49 per cent; boys looked at the mobile for 52 per cent of the time; girls, 41 per cent.) Four to five months later (according to a better controlled study), both boys and girls prefer to look at people than at objects, to the same degree.[26] In the second year of life differences do seem to emerge, but they are still rather subtle. A large recent study of nearly one hundred two-year-old children measured how long they played with a doll and a truck (among other toys), and how often they nurtured or manipulated the toys. About a third of the time a randomly chosen boy would play in a more "girlish" (or non-"boyish") way than a randomly chosen girl, both in terms of what toy they played with, and how they played with it.[27] And sometimes at this age kids play as long, or longer, with counter-stereotypical toys than with those that are supposedly "for them": like the fourteen-month-old boys in one study who played for about twice as long with a tea set as they did with a truck, a train, and motorcycles, put together (while the girls in this study spent as long with these "boy toys" as they did with dolls).[28]

So how do we get from this to the more markedly stereotypical toy preferences children come to develop? In keeping with the suggestions of cultural evolutionists, developmental psychologists describe young children as "gender detectives."[29] Children see that the category of sex is the primary way that we carve up the social world, and are driven to learn what it means to be male or female. Then once they come to understand their own sex, at about two to three years of age, this information takes on a motivational element: kids begin to "self-socialize" (sometimes to the chagrin of feminist parents). Presumably not coincidentally, this is the time period during which many boys start to shun pink, and many girls become especially drawn to it.[30] By just three years of age, when children are presented with other kids endorsing novel, gender-neutral objects and activities, they show "robust preferences" for those promoted by kids of the same sex.[31]

In fact, a recent study led by Cambridge University psychologist Melissa Hines suggests that at least part of the reason that girls with CAH have more boyish play interests is because they're less influenced by gender labels and gender modelling than are other children.[32] Four- to eleven-year-old matched control girls (and boys with and without CAH) preferred a gender-neutral toy that was presented either explicitly or implicitly as being "for them" (echoing findings from the 1970s and 1980s).[33] By contrast, girls with CAH were impervious to information that particular toys (like a xylophone or balloon) were "for girls," despite remembering that information just as well. This makes sense, given the somewhat weaker female gender identity of this population.[34] In my previous book, *Delusions of Gender*, I pointed out that studies of girls with CAH are done in ways that leave open the possibility that these girls aren't, in fact, drawn to some unidentified quality intrinsic to "boy toys" that appeals to their "masculinized" brains, but simply identify more than do girls without the condition to masculine activities, whatever those might be in a particular time, place, and culture.[35] Along similar lines, Barnard College sociomedical scientist Rebecca Jordan-Young points out that to understand these girls' more masculine preferences, we have to consider the psychosexual effects of the condition: girls are born with atypical or masculinized genitalia, they often undergo intensive medical and psychiatric observation or intervention, and have physical characteristics out of keeping with cultural ideals of feminine attractiveness.[36]

Certainly, as with novel and gender-neutral objects, children's interest in even counter-stereotypical toys can be piqued by seeing a child of the same sex play with it.[37] And more recent evidence points to the influence of the now-ubiquitous colour coding of gender. Psychologist Wang Wong, together with Melissa Hines, compared how long boys and girls played with a train and a doll, first when they were twenty to forty months old, and then again about half a year later.[38] At both ages, it's worth pointing out, girls played longer with the train than with the doll. (Draw whatever conclusions you will regarding the implications for the "naturalness" of child care as an occupation for women, compared with the much better remunerated occupation of mechanical engineer.) But the researchers' main interest was in whether children were influenced by the *colour* of the toys. Lo and behold, sex differences in toy preferences were smaller when children were presented with a pink train and a blue doll than when presented with the same toys in stereotypical colours. In fact, at the slightly older age, the same boys and girls showed moderate to large differences in the amount of time they spent playing with a blue train and a pink doll, but small and statistically indistinguishable amounts of time playing with a pink train and a blue doll.[39] Whatever role, if any, testosterone or other facets of biological sex play in girls' and boys' initial overlapping toy preferences (and there are other possible explanations), all of this is troublesome for the Testosterone Rex perspective. One doesn't expect a deeply biologically rooted, evolved sex-specific nature to be so contradictory and inconsistent in its expression, or to be so easily overridden by a quick paint job.

From birth, children encounter endless gender clues and hints in the real world: gender stereotypes transmitted in advertisements; encouraging or discouraging words, expressions, or body language from others; toy stores and packaging; movies; TV shows; the sex-segregation of adult social roles; and so on. Of course, these many influences don't impose themselves onto a blank slate: every child is different, with their own internal inclinations and understandings. Some influences will leave particular children untouched while affecting others. (Interestingly, it may be that children who have a stronger "lens of gender" may be especially susceptible to the influence of stereotypical information.)[40] Some gender messages will push in opposite directions, and no single influence is likely to be very large. But they accumulate. And they provide a potential explanation for how robust sex differences in toy preferences develop around the age that children develop a firm understanding of which side of the critical social divide of gender they belong. The gendered developmental system has achieved what prenatal testosterone can't.

This conclusion, by the way, is perfectly consistent with claims that back in our evolutionary past it was adaptive for women and men to have had very different roles: for women to care for children, and for men to handle spears and kill stuff. It's compatible with this being a common pattern across societies. And it's also perfectly reconcilable with things being very different now, and different again in the future.

As Paul Griffiths explains, it's well accepted in evolutionary biology that even adaptive traits that increase reproductive success can take different forms, depending on environmental conditions.[41] (Evolutionary Psychology, for instance, famously describes this in terms of a jukebox metaphor: various possible behavioural "tunes" are built into the genes, and which one gets "played" depends on circumstances.)[42] Just ask your nearest dung beetle. The male cichlid fish of Chapter 6 provide another striking and more dynamic example. Whether a male develops into a dominant fish—physically, behaviourally, and hormonally—depends on his social situation and real estate conditions. A fish placed in a tank with a smaller fish will become dominant, a fish without a breeding territory will remain subordinate, and hormones follow status. Or recall the female bush crickets from Chapter 1 that compete for males bearing nutrient-rich sperm packages when times are bad, but sit back and choose when the living is easy. Then there were the hedge sparrows, that allow the wildly variable sexual mores of their mating system to be set by, among other factors, the happenstance of the location of their breeding territories. These animals certainly seem to be behaving adaptively, but that behaviour clearly isn't fixed by their genes or nature. What we can conclude from these examples is that just because a particular kind of behaviour is adaptive in certain conditions doesn't mean it's fixed and will develop regardless.

But what about adaptations that are standard issue, that we see in a species regardless of environmental or social circumstances? Wouldn't these be locked into genetically inherited biology, to ensure they develop? Not necessarily. Recall the rat mothers of Chapter 4, that especially vigorously lick the anogenital region of their male pups. This strange phenomenon illustrates that natural selection is a frugal process that can, and does, draw on stable and reliable inputs from the developmental system, beyond the genes. Griffiths has another nice example: the ability of rhesus macaque monkeys to recognize emotional expressions and successfully navigate conflict. The development of these skills, despite being obviously highly adaptive, turn out to depend on social contact and interactions in infancy. But that's fine, because these are social experiences every young rhesus macaque monkey will reliably encounter in the normal course of events, generation after generation. As Griffiths points out, that rhesus macaques need a particular kind of early social input in order to develop these abilities "throw[s] no doubt whatever on the claim that these abilities in adult macaques are the result of adaptive evolution."[43] Indeed, in the rat case, the mother's licking contributes to the development of something as fundamentally adaptive as sexual behaviour.

What does all of this mean for ourselves, given the monumental ecological, technological, social, medical, and cultural changes that have taken place throughout human history? As John Dupré points out:

> Since the conditions under which contemporary brains develop are very different from the conditions under which human brains developed in the Stone Age, there is no reason to suppose that the outcome of that development was even approximately the same then as now.

This, he takes pains to point out, is not to say "that brains are blank slates developing with infinite plasticity in response to environmental variation." It simply takes seriously the role of the developmental system in development and evolution: the "brain [is] constructed by a variety of more or less stable and reliable resources including resources that are reliably reproduced by human cultures."[44]

So can even universal, adaptive traits be obliterated with a simple manipulation to the environment? Consider an experiment in which scientists selectively bred two lines of mice, one high in aggression, the other low. They achieved this by putting the young mice in social isolation after weaning, which increases aggressive tendencies in some. Then, mice for which this was the case—that were particularly combative in a staged encounter with another mouse four weeks later—were selectively bred together. Likewise for mice that were the least aggressive. After just seven generations of this selective breeding programme, the researchers successfully created two lines of mice that behaved in very different ways. When reared in isolation, mice bred to be thuggish were about six times more aggressive than the other group. After thirty-nine generations, the two lines had diverged even further. Aggression had therefore become a consistent, "adaptive" trait in the antagonistic line of mice (with the scientists acting as the hand of natural selection, increasing the reproductive success of the most aggressive mice). But here's the remarkable part. Despite a heritage of thirty-eight ancestors bred for aggression behind them, thuggish mice that were reared in a different way from their ancestors—with other mice, instead of in isolation—turned out *no more aggressive* than mice bred for generations to be gentle.[45] A simple but critical change to the developmental system eliminated a typical, "adaptive" trait.

Here's another example that some overworked mothers might find inspiring. We saw in Chapter 2 that being the one who produces the sperm doesn't dictate, by universal principle, that parenting is out of the portfolio. However, in the case of the rat (as with most mammals), the balance of trade-offs make it more adaptive for males to leave parenting to the mothers. This might tempt us to take it for granted that males, by virtue of their sex, therefore lack the capacity to care for pups. We might well assume that, through sexual selection, they lost or never acquired the biological capacity to parent: that it isn't "in" their genes, hormones, or neural circuits. That it isn't in their male nature. But bear in mind that one reliable feature of a male rat's developmental system is a female rat that does the child care. So what happens when a scientist, under controlled laboratory conditions, simulates a first-wave feminist rodent movement by placing males in cages with pups but no females? Before too long you will see the male "mothering" the infant, in much the same way that females do.[46]

However surprising these two examples may seem, they are perfectly compatible with modern evolutionary thinking—just not with how most of us are used to thinking about adaptations. When we say, for instance, that sex differences in children's toy preferences are "innate," we usually fold three different assumptions into that word, as Griffiths explains. First, we mean that boys' and girls' preferences reflect an evolutionary adaptation: girls like dolls because they are adapted to care for babies; boys like toy trucks because, well, trucks move, and so do spears and animals when you hunt. The second assumption we usually make when we say something is innate is that it's fixed. In the case of toys, we mean that neither feminist parenting nor gender-neutral marketing can eliminate those innate interests. And the third thing we often imply is that a preference for stereotypical toys is, if not universal, then at least typical of boyhood and girlhood. All this is what we mean, too, when we say that *boys will be boys*. Essentialist thinking leads us to bundle together these three biological properties: adaptiveness, fixity, and typicality, argues Griffiths. We tend to assume that if a behaviour or trait is an adaptation, then it must also be fixed and typical. Conversely, if it seems that a characteristic is typical (or universal), then it must also be fixed, and also probably an adaptation. This is why so much seems to hinge, politically and socially, on scientific questions like "Is it universal across societies for men to be higher in status?" and "Are men more promiscuous than women cross-culturally?"

Sometimes, certainly, these three biological properties *do* come together as a package: like the female and male human reproductive systems. The female human reproductive system is an essential trait of femaleness: it is adaptive; it develops in more or less the same way across a wide range of environmental, physical, social, and cultural conditions; and is highly typical (although not universal) in genetic females. But a well-accepted principle in developmental science is that adaptiveness, fixity, and typicality *don't* necessarily come together. That a trait checks off one of the three boxes doesn't mean it also checks both, or either, of the other two. Since, for instance, the development of adaptive traits relies on the entire developmental system, not just the genes, a relevant change in the external developmental system can change an adaptive behaviour: like the male rats that became parental when physically put in what would normally be the mother's place. That is, adaptive traits don't necessarily develop regardless of conditions. Nor are adaptations necessarily typical. Evolution can produce different forms of an adaptive trait: like the male dung beetles that can be either armoured and belligerent, or come in the hornless form that considers discretion to be the better part of valour. And behaviours can also be typical without being either an adaptation, or fixed. Even in a world in which all women wore dresses, we wouldn't want to say that dress-wearing was a developmentally fixed sex-specific adaptation.

This disentangling means that the answer to questions like "Were male promiscuity, risk taking, and competitiveness sexually selected adaptations for reproductive success?" simply don't have the implications for now and the future that we usually assume they do: that if the answer is "yes," then *boys will be boys*. But when we think in essentialist ways about social groups, the differences between them seem "large, unbridgeable, inevitable, unchangeable, and ordained by nature," as University of Melbourne psychologist Nick Haslam summarizes it.[47] Those who think in gender-essentialist ways are more likely to endorse the gender stereotypes that are the foundation of intended and unintended discrimination in the workplace.[48] They are more likely to feel negatively towards power-seeking women, relative to men.[49] They are more likely to allocate child care in a traditional way.[50] They are more likely to prefer that the husband earns more in a heterosexual marriage, and to expect to make traditional work-care trade-offs.[51] Women encouraged to take an essentialist view of gender become more vulnerable to "stereotype threat"—the reduction in performance and interest in traditionally masculine domains triggered by negative stereotypes about women.[52] Gender essentialist thinking makes men evaluate sex crimes more leniently,[53] and makes people less supportive of progressive gender policies and feel more comfortable with the status quo.[54]

That's why the evidence that sex hasn't "fixed" any behaviours as "essential" traits is so important. Instead, the genetic and hormonal components of sex collaborate with other parts of the developmental system, including our gender constructions. There have been massive shake-ups in that developmental system since the Pleistocene—laws, social welfare, taxation, medical advances, industrialization, and so on. And while the male and female reproductive systems have stayed the same across human history, as the developmental system has changed—whether through the introduction of contraception, equal opportunity legislation, paternity leave, or gender quotas—brains, hormones, behaviour, and roles have changed.

We already know that when this happens in a biggish way, the changes to gendered behaviour can be remarkable. In our postindustrial societies, reliable contraception and technology have made the physical differences between the sexes less important, and this has led to a rapid merging of sex roles, as Wendy Wood and Alice Eagly point out.[55] Women have stampeded into traditionally masculine roles like law, medicine, accounting, and management. Although there hasn't been a reciprocal rush by men towards traditionally feminine roles, like early education and nursing, this might be expected purely on the grounds of the unappealingly low status and pay of "women's work."[56] Or to take another example, as Jordan-Young documents in *Brain Storm*, only thirty or forty years ago scientists categorized so many sexual behaviours as distinctly masculine—the initiation of sex, intense physical desire, masturbation, erotic dreams, arousal to narratives—that it was "scarcely an overstatement to suggest that sexuality itself was seen as a masculine trait."[57] Female sexual imagination was restricted to "wedding fantasies" (presumably not of an "Ooh, *Reverend!*" variety). As for the contemporary $1 billion U.S. market for vibrators,[58] to the psychobiologists of the time, this would presumably have indicated an epidemic of abnormal female sexuality on a catastrophic scale. "From this side of the sexual revolutions of the 20th century, it is easy to lose track of just how much has changed, and how rapidly,"[59] Jordan-Young observes.

What does this mean for the aspiration to see a more balanced society, from more boys playing with dolls and more dads caring for kids, to more women in science and senior leadership roles? As Dalhousie University philosopher Letitia Meynell puts it:

> Biologically speaking, our actions and dispositions are developed and could have been otherwise, given the right mix of developmental inputs at various points in our lives. If one wants to change the distribution of a given trait in a population, *the task is not to overcome nature but to rearrange the developmental system.*[60]

While this is a rightly optimistic message, rearranging the developmental system is no trivial task. Ironically, the rich, stable cultural inheritance that enables us to be so adaptably diverse as a species is also a heavy counterweight to change. If you want a male rat to take care of a baby rat, you can just pop him in a cage with one. Rearranging gender in the human developmental system involves the reconstruction of the social structures, values, norms, expectations, schemas, and beliefs that penetrate our minds, interactions, and institutions, and that influence, interact, and become entangled with our biology. There's a reason they're called "social constructions" rather than, say, "social Legos." Social constructions are robustly built: you can pull out bricks here and there, but the others continue to hold everything in place. They're not easily torn apart and reconstructed in new ways.

Take domestic violence. What makes a person, usually a man, more likely to assault a partner or former partner? Experts point the finger at a dauntingly long list of influences, including rigid gender stereotypes that tightly circumscribe appropriate female roles and responsibilities, hypermasculine norms, societal excuse making for violence towards female partners, lack of perpetrator accountability, many women's economic dependence on their male partners, a society that places females lower in status than males, and government's low financial and political investment in the problem.[61] That's a lot of collective rearranging to be done if we're to reduce the number of men harming women. How much simpler a problem it would be if violent men simply had too much testosterone.

So how should we think about those gender-coded toy aisles now—those pink and blue plastic safety knives being sold at the school market?

A year on, in the lead-up to the next Christmas, Australian senator Waters drew links a second time between the rigid gender stereotypes promoted by sex-segregated toy marketing, and seemingly far-removed social issues, like the gender pay gap and domestic violence.[62] More scorn was poured. But now think about gendered toy marketing not as boy versus girl nature made manifest, but part of the developmental system. At the very time children are laying down cultural meanings and norms in their minds, gendered marketing emphasizes sex as a critically important social divide.[63] That booth seller at the local school market overlooked everything her two small customers had in common—their family background, their close age, their ethnicity, the shared fortune of a parent who doesn't see sliced fingers as an inevitable and important childhood learning experience—and instead emphasized one thing that was different, their genitals. And while the colour coding of any toy or product sends this message, when those gender cues are also linked to stereotypical *kinds* of toys and products, it also serves the no less important purpose of reinforcing stereotypes of males as "bad but bold" masters of the world and females as "wonderful but weaker" carers.[64] These gender stereotypes operate throughout life both as expectations about the characteristics men and women have, and as gender norms dictating double standards for how women and men should behave, influencing people's interests, self-concept, performance, and beliefs about capabilities in gendered domains. These gender stereotypes and norms are also the foundation of both conscious and unconscious forms of sex discrimination, like biased evaluations of performance and potential, and social and economic backlash against people whose behaviour isn't in line with them.[65]

Gender stereotypes and norms can certainly harm and constrain boys and men too. But gender is a hierarchy. The higher prestige of males and masculinity is, some have speculated, why significant numbers of girls in middle childhood start to shun the "girl" toys and activities they have supposedly evolved to prefer, in order to become "one of the boys," while there is a conspicuous absence of boys hoping to become "one of the girls."[66] And since traditionally masculine occupations and roles are generally associated with more prestige and better pay than equivalently skilled feminine roles, gender stereotypes and norms are particularly harmful to women financially and professionally. Ironically, unconscious gender bias is now considered such an obstacle to the fair promotion and retention of women that organizations routinely invest considerable time and money in training to reduce it—yet we vigorously sow the seeds of it in our children from the moment they are born.

So what do we want? Do we want a society that genuinely values equal opportunity for development, employment, economic security, safety, and respect, regardless of sex? If so, there's a glaring contradiction with the messages some marketers are sending to children. As psychologists Sheila Cunningham and Neil Macrae point out, the colour coding of toys "seems at odds with the egalitarian goals that feature so prominently in contemporary society."[67]

Toy marketing is obviously just one strand of many that weave gender through the developmental system. No single factor is overwhelmingly important in creating sex inequalities. Every influence is modest, made up of countless small instances of its kind. That's why everything—a doll packaged in pink, a sexist joke, a male-only expert panel—can seem trivial, of intangible effect. But that's exactly why calling out even seemingly minor points of sexism matters. It all adds up, and if no one sweats the small stuff, the big stuff will never change. Senior leaders obviously enjoy the most power to create change—whether through implementing targets and quotas, pay gap audits, more generous paternity leave, rooting out sexual harassment, or rethinking media portrayals—but everyone else, and there are a lot of us, can play a part: complaining about doll and kitchen toy aisles labelled "for girls" and science kits labelled "for boys"; petitioning for women's achievements to also be honoured on paper money; even asking for a plastic knife in the "wrong" colour. How efforts and money are best recruited to achieve a social goal, and how much regulation should be invoked, are certainly legitimate questions for debate. But if we, as a society, say that we are for equality between the sexes, then when someone has the courage to speak up and ask for change, for something better, fairer, less sexist, or more respectful, they don't deserve to be shot down with charges of insanity, overreaction, or political correctness gone mad.

So, again, what do we want?

PEOPLE HAVE DIFFERENT REASONS for wanting greater equality between the sexes. Some people want fewer women assaulted or killed by their partners. Some want to close the yawning gap in retirement savings that puts disproportionate numbers of women in poverty in their senior years. Some want greater sex equality in their organizations because of research suggesting beneficial effects for productivity and profit. Some people want mothers and fathers to share more equally in caring for children so that the next generation reaps the benefits of involved, caring fathers and happier parents. Some people want an easier journey for loved ones with identities, bodies, or both, that fall in-between the too-neat male versus female binary. Some want it to become easier for people to pursue and fulfil counter-stereotypical ambitions. Others want to stem the leak of talented, highly educated, and expensively trained women lost in professional pipelines. Some want to see households headed by single mothers lifted out of hardship or poverty. Some want more equal political representation, so that girls' and women's interests are more equally served in government policy. Some people are also for sex equality because of a suite of benefits for men: from lessening of pressure to live up to demanding and sometimes dangerous hypermasculine norms, to an easing of the burden and stresses of being the primary breadwinner. Some hope it will bring a liberating expansion of the definition of male success into the parts of human existence beyond work, wealth, and sexual conquest. Some go even further, and hope that thinking of qualities, roles, and responsibilities as *human*, rather than as feminine or masculine, will transform the world of work, to the benefit of everyone. Others think that greater sex equality is probably a mixed bag for men, but that we should try for it anyhow because it's just fairer and nicer when power, wealth, and status are more equally shared.

And some people think that sex equality is a lovely idea in principle, but that Testosterone Rex fiercely blocks the path to this better place. Why? Because men are from Mars and women are from Venus, a woman can't be like a man, and boys will be boys.

But I've never heard anyone admit to holding the following view: *Look, I agree, it's not very fair. Nor is it decreed by nature, so we could change things a lot, if we wanted. But we've had sex inequality for thousands of years and I kind of like it. So how about we just keep things as they are?*

Apparently we're all for sex equality then. So what now?

We can decide it's too much trouble, and settle for a half-changed world. Alternatively, we can continue with our polite, undemanding panel discussions about gender equality, our good intentions and gentle tinkering, and patiently wait out the fifty to one hundred or so years it's regularly predicted to take to achieve parity in the workplace. But if neither of these options is appealing, then maybe it's time to be less polite and more disruptive; like the first- and second-wave feminists. They weren't always popular, it's true. But look at what they achieved by not asking nicely.[68] Words are nice, but often deeds work better.

Which of these directions we prefer is up to us: it's a question for our values, not science. But that evolving science is showing that one time-honoured option is no longer available to us. It's time to stop blaming Testosterone Rex, because that king is dead.

South Dublin Libraries
www.southdublinlibraries.ie

ACKNOWLEDGEMENTS

THANK YOU TO THE MANY PROFESSIONALS WHO IN VARIOUS WAYS helped to bring this book into being. I thank in particular my wonderful and tireless agent, Barbara Lowenstein, and her team at Lowenstein Associates. I am also very grateful to everyone who assisted at Norton, most especially my editor, Amy Cherry, for her thoughtful and careful work on the manuscript, her encouragement, and her patience, and Remy Cawley for her efficient and good-natured assistance. Many thanks also to my meticulous copyeditor, Nina Hnatov. A big thank-you, too, to the wonderful team at Icon Books, most especially Kiera Jamison. I'm also immensely grateful to the many academics who gave their time and expertise so generously and willingly to comment on one or more chapters, or early guidance: Elizabeth Adkins-Regan, John Dupré, Anne Fausto-Sterling, Agustín Fuentes, Martha Hickey, Daphna Joel, Julie Nelson, Elise Payzan-LeNestour, Sari van Anders, and Bill von Hippel. I'd also like to especially thank three treasured colleagues at the University of Melbourne—Mark Elgar, Nick Haslam, and Carsten Murawski—who additionally provided moral support and obliging responses to strange book-related questions over the years of its writing, as well as substantial and invaluable feedback on the manuscript. Progress on this book was greatly assisted by a Future Fellowship from the Australian Research Council, held at the Melbourne School of Psychological Sciences, University of Melbourne, and by the support of the Melbourne Business School, and the Centre for Ethical Leadership at Ormond College, University of Melbourne. I am also grateful for the support of the Women's Leadership Institute Australia Research Fellowship.

Some passages in the book draw on previously published work. Chapter 3 includes material that was previously published in very similar form in Fine, C. (November, 2012). The vagina dialogues: When it comes to libido, testosterone is overrated. *The Monthly*. Chapter 4 includes lines that were previously published in Joel, D., & Fine, C. (December 1, 2015). It's time to celebrate the fact that there are many ways to be male and female. *The Guardian*. Thank you to John van Tiggelen and Ian Sample, respectively, for engaging with this work. I am indebted to colleagues with whom I have coauthored work that contributed to some of the thinking and ideas presented in this book. Arguments in Chapter 8 were developed from Fine, C., & Duke, R. (2015). Expanding the role of gender essentialism in the single-sex-education debate: A commentary on Liben. *Sex Roles*, 72(9), 427–433. Chapter 8 was also strongly informed by work coauthored with Emma Rush, in Fine, C., & Rush, E. (2016). "Why does all the girls have to buy pink stuff?" The ethics and science of the gendered toy marketing debate. *Journal of Business Ethics*, doi:10.1007/S10551-016-3080-3. Some material in this chapter was presented as part of the third annual Alan Saunders Lecture, at the ABC Ultimo Centre, Sydney, July 7, 2015, presented by Radio National's *The Philosopher's Zone* and the Australasian Association of Philosophy.

Four colleagues—Daphna Joel, Rebecca Jordan-Young, Anelis Kaiser, and Gina Rippon—have played a critical role in developing my thinking about scientific models of sexual differentiation, and how to study sex/gender in humans, in part through the following works coauthored with them: Fine, C., Jordan-Young, R., Kaiser, A., & Rippon, G. (2013). Plasticity, plasticity, plasticity . . . and the rigid problem of sex. *Trends in Cognitive Sciences, 17*(11), 550–551; Rippon, G., Jordan-Young, R., Kaiser, A., & Fine, C. (2014). Recommendations for sex/gender neuroimaging research: Key principles and implications for research design, analysis, and interpretation. *Frontiers in Human Neuroscience, 8*, 650; Fine, C., Joel, D., Jordan-Young, R., Kaiser, A., & Rippon, G. (December 15, 2014). Why males ≠ Corvettes, females ≠ Volvos, and scientific criticism ≠ ideology. *Cerebrum*. These four colleagues have also been a wonderful source of intellectual and personal support throughout the writing of this book.

Last, I am indebted to Richard Francis for this book's title. As noted in Chapter 6, Francis used the term "Testosterone Rex" in his book *Why Men Won't Ask for Directions: The Seductions of Sociobiology.*

This is my third book, and it doesn't seem to get any easier. As usual, my thanks to Russell for being such an excellent and devoted father. Many thanks, too, to my writer-friends, Simon Caterson, Monica Dux, Christine Kenneally, and Anne Manne, whose talents extend beyond writing to also knowing exactly how best to encourage other writers. I give my heartfelt thanks, as always, to my mother, Anne Fine, for her editorial and emotional assistance throughout. Finally, my deepest thanks and appreciation to C-Rex, for his infinitely patient encouragement and support, and unflagging interest and confidence in this book. It would have been enough just to have put up with me.

NOTES

EPIGRAPH

1. TEDxEuston, April 12, 2013.

INTRODUCING TESTOSTERONE REX

1. Haslam, N., Rothschild, L., & Ernst, D. (2000). Essentialist beliefs about social categories. *British Journal of Social Psychology, 39*, 113–127. This study of Northern American students found that gender is a strongly essentialized social category, particularly with respect to being seen as being a "natural kind"—that is, being natural, fixed, invariant across time and place, and discrete (that is, with a sharply defined category boundary).
2. Dupré, J. (forthcoming). A postgenomic perspective on sex and gender. In D. L. Smith (Ed.), *How biology shapes philosophy: New foundations for naturalism*. Cambridge, UK: Cambridge University Press.
3. Charles Darwin (pp. 245–255) defined sexual selection as arising from "the advantage which certain individuals have over other individuals of the same sex and species, in exclusive relation to reproduction." Darwin, C. (1871). *The descent of man, and selection in relation to sex*. London: John Murray. Cited on p. 10001 of Jones, A. G., & Ratterman, N. L. (2009). Mate choice and sexual selection: What have we learned since Darwin? *Proceedings of the National Academy of Sciences, 106*(Suppl. 1), 10001–10008. Jones and Ratterman describe this as being highly similar or identical to contemporary definitions of sexual selection.

4. See, for example, Gangestad, S. W., & Thornhill, R. (1998). Menstrual cycle variation in women's preferences for the scent of symmetrical men. *Proceedings of the Royal Society of London B: Biological Sciences, 265*(1399), 927–933; Pillsworth, E. G., & Haselton, M. G. (2006). Women's sexual strategies: The evolution of long-term bonds and extrapair sex. *Annual Review of Sex Research, 17*(1), 59–100.
5. Quoted in Elgot, J. (July 12, 2014). "We can't force girls to like science," says Glasgow University academic Dr. Gijsbert Stoet. *Huffington Post*. Retrieved from http://www.huffingtonpost.co.uk/2014/07/12/girls-science_n_5580119.html on August 7, 2015.
6. Yamabiko, M. (May 9, 2015). Women in F1 a question of brawn or brain? *www.crash.net*. Retrieved from http://www.crash.net/f1/feature/218544/1/max-yamabiko-women-in-f1-a-question-of-brawn-or-brain.html on August 7, 2015.
7. Cahill, L. (April 1, 2014). Equal ≠ the same: Sex differences in the human brain. *Cerebrum*.
8. This is a point made by Daphna Joel. For example, Joel, D. (2014). Sex, gender, and brain: A problem of conceptualization. In S. Schmitz & G. Höppner (Eds.), *Gendered neurocultures: Feminist and queer perspectives on current brain discourses* (pp. 169–186). Austria: University of Vienna/Zaglossus.
9. Paul Irwing, quoted in Rettner, R. (January 4, 2012). Men and women's personalities: Worlds apart, or not so different? *Live Science*. Retrieved from http://www.livescience.com/36066-men-women-personality-differences.html on February 10, 2016.
10. Cahill (2014), ibid.
11. As noted in Fine, C., Joel, D., Jordan-Young, R. M., Kaiser, A., & Rippon, G. (December 15, 2014). Why males ≠ Corvettes, females ≠ Volvos, and scientific criticism ≠ ideology. *Cerebrum*.
12. For example, Auster, C., & Mansbach, C. (2012). The gender marketing of toys: An analysis of color and type of toy on the Disney store website. *Sex Roles, 67*(7–8), 375–388; Blakemore, J., & Centers, R. (2005). Characteristics of boys' and girls' toys. *Sex Roles, 53*(9/10), 619–633.
13. Gray, J. (2012). *Men are from Mars, women are from Venus: The classic guide to understanding the opposite sex*. New York: HarperCollins.
14. Farrel, B., & Farrel, P. (2001). *Men are like waffles—women are like spaghetti; Understanding and delighting in your differences*. Eugene, OR: Harvest House.
15. Pease, A., & Pease, B. (2010). *Why men want sex and women need love: Solving the mystery of attraction*. London: Orion Books.
16. Pease, A., & Pease, B. (2001). *Why men don't listen and women can't read maps: How we're different and what to do about it*. London: Orion Books.
17. Moir, A., & Moir, B. (2003). *Why men don't iron: The fascinating and unalterable differences between men and women*. New York: Citadel Press/Kensington.
18. Moss, G. (2014). *Why men like straight lines and women like polka dots: Gender and visual psychology*. Alresford, UK: Psyche Books.
19. Shambaugh, R. (2013). *Make room for her: Why companies need an integrated*

leadership model to achieve extraordinary results. New York: McGraw-Hill. Quoted from front flap.

20. Gray, J., & Annis, B. (2013). *Work with me: The eight blind spots between men and women in business*. New York: St. Martin's Press.
21. Adams, S. (April 26, 2013). Eight blind spots between the sexes at work. *Forbes*. Retrieved from http://www.forbes.com/sites/susanadams/2013/04/26/8-blind-spots-between-the-sexes-at-work/ on April 28, 2013; Girl talk. (April 13, 2013). *The Economist*. Retrieved from http://www.economist.com/news/business-books-quarterly/21576073-working-women-today-have-it-better-ever-few-agree-how on April 15, 2013.
22. Herbert, J. (2015). *Testosterone: Sex, power, and the will to win*. Oxford, UK: Oxford University Press. Quoted on p. 194.
23. Herbert suggests that testosterone plays an important role in women in relation to sexuality, and acknowledges a relative lack of knowledge with respect to its role in females.
24. This implicit or explicit assumption is described by van Anders, S. M. (2013). Beyond masculinity: Testosterone, gender/sex, and human social behavior in a comparative context. *Frontiers in Neuroendocrinology, 34*(3), 198–210.
25. Alexander, R. D. (1979). *Darwinism and human affairs*. Seattle: University of Washington Press. Quoted on p. 241.
26. Hewlett, S. (January 7, 2009). Too much testosterone on Wall Street? *Harvard Business Review Blogs*. Retrieved from http://blogs.hbr.org/2009/01/too-much-testosterone-on-wall/ on April 15, 2010.
27. Dupré, J. (1993). Scientism, sexism and sociobiology: One more link in the chain. *Behavioral and Brain Sciences, 16*(2), 292. Quoted on p. 292.
28. See illuminating discussion on this point in Wilson, D. S., Dietrich, E., & Clark, A. B. (2003). On the inappropriate use of the naturalistic fallacy in evolutionary psychology. *Biology and Philosophy, 18*(5), 669–681.
29. Wilson et al. (2003), ibid. See also Dupré, J. (2003). *Darwin's legacy: What evolution means today*. Oxford, UK: Oxford University Press; Kennett, J. (2011). Science and normative authority. *Philosophical Explorations, 14*(3), 229–235; and Meynell, L. (2008). The power and promise of developmental systems theory. *Les Ateliers de L'Ethique, 3*(2), 88–103.
30. Kennett (2011), ibid. Quoted on p. 229.
31. Quoted in Elgot (2014), ibid.
32. Browne, K. R. (2012). Mind which gap? The selective concern over statistical sex disparities. *Florida International University Law Review, 8*, 271–286. Quoted on p. 284, footnoted references that follow excluded.
33. Hoffman, M., & Yoeli, E. (Winter, 2013). The risks of avoiding a debate on gender differences. *Rady Business Journal*.
34. Tom Knox, quoted while Chairman of DLKW Lowe in the Marketing Society Forum. (March 7, 2014). Should all marketing to children be gender-neutral? *Marketing* (March 7). Retrieved from http://m.campaignlive.co.uk/article/1283685/marketing-children-gender-neutral on September 8, 2014.

35. Tony Abbott, speaking as prime minister of Australia, quoted in Dearden (December 2, 2014). Tony Abbott says campaigners against gendered toys should "let boys be boys and girls be girls." *The Independent*, (December 2), Retrieved from http://www.independent.co.uk/news/world/australasia/tony-abbott-says-campaigners-against-gendered-toys-should-let-boys-be-boys-and-girls-be-girls-9897135.html on April 27, 2015.
36. Liben, L. (2015). Probability values and human values in evaluating single-sex education. *Sex Roles*, *72*(9–10), 401–426. Quoted on p. 415. Liben herself does not hold a gender-essentialist view, and notes that from this perspective, sex is seen as so fundamental "that other potentially important human characteristics may pale in comparison, and thus may attract relatively little attention."
37. For example, Halsam et al. (2000), ibid.; Haslam, N., Rothschild, L., & Ernst, D. (2000). Are essentialist beliefs associated with prejudice? *British Journal of Social Psychology*, *41*, 87–100; Rothbart, M., & Taylor, M. (1992). Category labels and social reality: Do we view social categories as natural kinds? In G. Semin & K. Fiedler (Eds.), *Language, interaction and social cognition* (pp. 11–36). Thousand Oaks, CA: Sage.
38. See Griffiths, P. E. (2002). What is innateness? *The Monist*, *85*(1), 70–85.
39. John Coates, quoted in Adams, T. (June, 19, 2011). Testosterone and high finance do not mix: So bring on the women. *The Guardian*. Retrieved from http://www.theguardian.com/world/2011/jun/19/neuroeconomics-women-city-financial-crash on February 20, 2014.

A NOTE ABOUT TERMINOLOGY

1. The first use of the word "gender" as dissociable from biological sex is attributed to John Money, in relation to the concepts of "gender identity" and "gender role" in 1955. However, Ann Oakley's 1972 book *Sex, Gender, and Society* seems to be the first publication to use the term to distinguish biological sex from cultural gender.
2. See Haig, D. (2004). The inexorable rise of gender and the decline of sex: Social change in academic titles, 1945–2001. *Archives of Sexual Behavior*, *33*(2), 87–96. However, my impression is that in the last decade or so, psychologists who emphasize biological contributions to differences between males and females tend to use "sex," while those who emphasize social contributions use "gender."
3. As Haig (2004, p. 87), ibid., puts it, "The use of gender has tended to expand to encompass the biological, and a sex/gender distinction is now only fitfully observed."
4. For example, Kaiser, A. (2012). Re-conceptualizing "sex" and "gender" in the human brain. *Journal of Psychology*, *220*(2), 130–136.
5. In addition to the disadvantage of imprecision, the word "promiscuous" in everyday language can also be taken to imply a lack of discrimination when it comes to choice of sexual partners. See Elgar, M. A., Jones, T. M., & McNamara, K. B. (2013). Promiscuous words. *Frontiers in Zoology*, *10*(1), 66.

6. University of Minnesota behavioural ecologist and evolutionary biologist Marlene Zuk has a wonderful description of the moral reactions of her students on learning that some bird mating systems aren't as monogamous as once thought. Zuk, M. (2002). *Sexual selections: What we can and can't learn about sex from animals*. Berkeley: University of California Press.

CHAPTER 1: FLIES OF FANCY

1. Extremely useful accounts, drawn on extensively here, are provided in Dewsbury, D. (2005). The Darwin-Bateman paradigm in historical context. *Integrative and Comparative Biology, 45*(5), 831–837; Tang-Martínez, Z. (2010). Bateman's principles: Original experiment and modern data for and against. In M. Breed & J. Moore (Eds.), *Encyclopedia of animal behavior* (Vol. 1, pp. 166–176). Oxford, UK: Academic Press; Hrdy, S. B. (1986). Empathy, polyandry, and the myth of the coy female. In R. Bleier (Ed.), *Feminist approaches to science* (pp. 119–146). New York: Pergamon Press; Tang-Martínez, Z. (2016). Re-thinking Bateman's principles: Challenging persistent myths of sexually-reluctant females and promiscuous males. *Journal of Sex Research, 53*(4–5), 532–559. Many examples drawn on in this chapter were originally cited in this critical review article.
2. Knight, J. (2002). Sexual stereotypes. *Nature, 415*, 254–256. Quoted on p. 254.
3. Darwin, C. (1871). *The descent of man, and selection in relation to sex*. London: John Murray. Quoted on p. 272.
4. Bateman, A. J. (1948). Intra-sexual selection in Drosophila. *Heredity, 2*(3), 349–368.
5. See Dewsbury (2005), ibid.
6. Trivers, R. L. (1972). Parental investment and sexual selection. In B. Campbell (Ed.), *Sexual selection and the descent of man* (pp. 136–179). Chicago: Aldine.
7. Tang-Martínez (2010), ibid. Both quotations from p. 167.
8. For instance, Patricia Gowaty is the editor of the 1997 book *Feminism and Evolutionary Biology: Boundaries, Intersections, and Frontiers* as well as other scholarly publications in this area.
9. Snyder, B. F., & Gowaty, P. A. (2007). A reappraisal of Bateman's classic study of intrasexual selection. *Evolution, 61*(11), 2457–2468. Quoted on pp. 2458 and 2457, respectively.
10. In fact, a later attempt to replicate Bateman's study (by Patricia Gowaty and her colleagues) challenged Bateman's assumption that there would be a quarter-share each of offspring with a maternal mutation only, paternal mutation only, a double mutation, and no mutation. Double-mutation flies, in particular, were particularly unlikely to survive. Gowaty, P. A., Kim, Y.-K., & Anderson, W. W. (2012). No evidence of sexual selection in a repetition of Bateman's classic study of *Drosophila melanogaster. Proceedings of the National Academy of Sciences, 109*(29), 11740–11745.
11. Snyder & Gowaty (2007), ibid.
12. Tang-Martínez, Z., & Ryder, T. B. (2005). The problem with paradigms:

Bateman's worldview as a case study. *Integrative and Comparative Biology, 45*(5), 821–830. They also noted that the mating behaviour of inbred strains may not have been representative of that seen in normal animals. Additionally, a longer experiment might have yielded different findings, since females can store sperm for a number of days, and only reach sexual maturity at about four days of age (compared with one day for males).

13. Tang-Martínez & Ryder (2005), ibid. Quoted on p. 821.
14. Snyder & Gowaty (2007), ibid., report that "Apparently, Bateman's sole rationale for graphing these series separately was that '. . .series 5 and 6 differed somewhat from the rest' [Bateman, 1948, p. 361]. This was neither a legitimate nor an a priori justification. Bateman goes on to note that series 5 and 6 were different from series 1–4 because the flies in series 5 and 6 were derived from crosses with inbred strains. However, series 4 was derived from inbred lines in a similar way, and, all six series differed in important ways including the number of flies in each population." Quoted on p. 2463.
15. Tang-Martínez (2010), ibid. Quoted on p. 168.
16. Synder & Gowaty (2007), ibid. Quoted on p. 2463.
17. See Tang-Martínez (2010), ibid.
18. See Table 1 in Gerlach et al. (2012), with examples from invertebrates, birds, fish, amphibians, reptiles, and mammals, for which female reproductive success is positively associated with multiple mating. Gerlach, N. M., McGlothlin, J. W., Parker, P. G., & Ketterson, E. D. (2012). Reinterpreting Bateman gradients: Multiple mating and selection in both sexes of a songbird species. *Behavioral Ecology, 23*(5), 1078–1088.
19. Schulte-Hostedde, A. I., Millar, J. S., & Gibbs, H. L. (2004). Sexual selection and mating patterns in a mammal with female-biased sexual size dimorphism. *Behavioral Ecology, 15*(2), 351–356; Williams, R. N., & DeWoody, J. A. (2009). Reproductive success and sexual selection in wild eastern tiger salamanders (*Ambystoma t. tigrinum*). *Evolutionary Biology, 36*(2), 201–213. Examples provided in Tang-Martínez (2016), ibid.
20. Imhof, M., Harr, B., Brem, G., & Schlötterer, C. (1998). Multiple mating in wild *Drosophila melanogaster* revisited by microsatellite analysis. *Molecular Ecology, 7*(7), 915–917.
21. Clapham, P. J., & Palsbøll, P. J. (1997). Molecular analysis of paternity shows promiscuous mating in female humpback whales (*Megaptera novaeangliae*, Borowski). *Proceedings of the Royal Society of London B: Biological Sciences, 264*(1378), 95–98.
22. Soltis, J. (2002). Do primate females gain nonprocreative benefits by mating with multiple males? Theoretical and empirical considerations. *Evolutionary Anthropology, 11*(5), 187–197. Quoted on p. 187.
23. See Chapter 4 of Zuk, M. (2002). *Sexual selections: What we can and can't learn about sex from animals*. Berkeley: University of California Press.
24. Lanctot, R. B., Scribner, K. T., Kempenaers, B., & Weatherhead, P. J. (1997). Lekking without a paradox in the buff-breasted sandpiper. *The American Naturalist, 149*(6), 1051–1070.

25. Lanctot et al. (1997), ibid. Quoted on p. 1059. As the researchers point out, these findings help to explain an apparent paradox of leks, which is why females should continue to exercise choice, despite (presumably) minimal genetic variation in male traits (due to almost all females breeding with the same male).
26. Hrdy (1986), ibid. See also citations in, for example, Tang-Martínez & Ryder (2005), ibid.
27. Hrdy (1986), ibid. Quoted on p. 135.
28. See p. 27 of Fuentes, A. (2012). *Race, monogamy, and other lies they told you: Busting myths about human nature.* Berkeley: University of California Press.
29. Hrdy (1986), ibid. Quoted on p. 137.
30. Brief overview provided in Knight (2002), ibid. For primates, see discussion in Soltis (2002), ibid. For a discussion of the potential genetic benefits of multiple matings for females, see Jennions, M., & Petrie, M. (2000). Why do females mate multiply? A review of the genetic benefits. *Biological Reviews, 75*(1), 21–64.
31. A point made by Hrdy (1986), ibid.
32. This latter possibility will be obscured by shorter-term studies as Hrdy (1986), ibid., points out. See also Stockley, P., & Bro-Jørgensen, J. (2011). Female competition and its evolutionary consequences in mammals. *Biological Reviews, 86*(2), 341–366.
33. See Table 1 in Stockley & Bro-Jørgensen (2011), ibid., p. 345.
34. Stockley & Bro-Jørgensen (2011), ibid. Quoted on p. 344. They note, however, that not all studies have found this effect, although in some cases this may have been because the populations under study were particularly well resourced.
35. For example, Blomquist, G. (2009). Environmental and genetic causes of maturational differences among rhesus macaque matrilines. *Behavioral Ecology and Sociobiology, 63*(9), 1345–1352.
36. A point made by both Hrdy (1986), ibid., and Stockley & Bro-Jørgensen (2011), ibid.
37. This point is made by a number of authors, including Baylis, J. (1981). The evolution of parental care in fishes, with reference to Darwin's rule of male sexual selection. *Environmental Biology of Fishes, 6*(2), 223–251; Dewsbury, D. (1982). Ejaculate cost and male choice. *The American Naturalist, 119*(5), 601–630; Dewsbury (2005), ibid.; and Tang-Martínez & Ryder (2005), ibid.
38. With considerable variability: Cooper, T. G., Noonan, E., von Eckardstein, S., Auger, J., Baker, H. W., Behre, H. M., et al. (2010). World Health Organization reference values for human semen characteristics. *Human Reproduction Update, 16*(3), 231–245.
39. Tang-Martínez & Ryder (2005), ibid. Quoted on p. 824.
40. Michalik, P., & Rittschof, C. C. (2011). A comparative analysis of the morphology and evolution of permanent sperm depletion in spiders. *PloS One, 6*(1), e16014. Cited in Tang-Martínez (2016), ibid.
41. A point made by Dewsbury (1982), ibid.

42. See Dewsbury (1982), ibid.
43. Tang-Martínez (2010), ibid. Quoted on p. 174.
44. For example, see Table 3, p. 318, of Wedell, N., Gage, M. J. G., & Parker, G. A. (2002). Sperm competition, male prudence and sperm-limited females. *Trends in Ecology and Evolution, 17*(7), 313–320.
45. Renfree, M. (1992). Diapausing dilemmas, sexual stress and mating madness in marsupials. In K. Sheppard, J. Boubilik, & J. Funder (Eds.), *Stress and reproduction* (pp. 347–360). New York: Raven Press. Both castrated males in the field, and those prevented from mating in the lab, survive substantially beyond the typical life span.
46. Elgar, M. (personal communication on August 6, 2015).
47. Alavi, Y., Elgar, M. A., & Jones, T. M. (2016). Male mating success and the effect of mating history on ejaculate traits in a facultatively parthenogenic insect (*Extatosoma tiaratum*). *Ethology, 122*, 1–8.
48. August, C. J. (1971). The rôle of male and female pheromones in the mating behaviour of *Tenebrio molitor. Journal of Insect Physiology, 17*(4), 739–751; Gwynne, D. T. (1981). Sexual difference theory: Mormon crickets show role reversal in mate choice. *Science, 213*(4509), 779–780; Pinxten, R., & Eens, M. (1997). Copulation and mate-guarding patterns in polygynous European starlings. *Animal Behaviour, 54*(1), 45–58.
49. Gowaty, P. A., Steinichen, R., & Anderson, W. W. (2003). Indiscriminate females and choosy males: Within- and between-species variation in *Drosophila. Evolution, 57*(9), 2037–2045.
50. Wade, M., & Shuster, S. (2002). The evolution of parental care in the context of sexual selection: A critical reassessment of parental investment theory. *The American Naturalist, 160*(3), 285–292. See Kokko, H., & Jennions, M. (2003). It takes two to tango. *Trends in Ecology and Evolution, 18*(3), 103–104.
51. Kokko, H., & Jennions, M. D. (2008). Parental investment, sexual selection and sex ratios. *Journal of Evolutionary Biology, 21*(4), 919–948. Quoted on p. 926.
52. Emlen, D. J. (1997). Alternative reproductive tactics and male-dimorphism in the horned beetle *Onthophagus acuminatus* (Coleoptera: Scarabaeidae). *Behavioral Ecology and Sociobiology, 41*(5), 335–341. With thanks to John Dupré for alerting me to this example.
53. Drea, C. M. (2005). Bateman revisited: The reproductive tactics of female primates. *Integrative and Comparative Biology, 45*(5), 915–923. Quoted on p. 920, references removed. Nor does parental care seem to be strongly tied to monogamy, a system in which males would have more certainty of paternity. See also Wright, P. C. (1990). Patterns of paternal care in primates. *International Journal of Primatology, 11*(2), 89–102.
54. Indeed, some biologists have argued that "sex role" is no longer a useful concept. For example, Ah-King, M., & Ahnesjö, I. (2013). The "sex role" concept: An overview and evaluation. *Evolutionary Biology, 40*(4), 461–470; Roughgarden, J. (2004). *Evolution's rainbow: Diversity, gender, and sexuality in nature and people.* Berkeley: University of California Press.

55. Gwynne, D. T., & Simmons, L. W. (1990). Experimental reversal of courtship roles in an insect. *Nature, 346*(6280), 172–174.
56. Forsgren, E., Amundsen, T., Borg, A. A., & Bjelvenmark, J. (2004). Unusually dynamic sex roles in a fish. *Nature, 429*(6991), 551–554. Quoted on p. 553.
57. Davies, N. (1992). *Dunnock behaviour and social evolution*. Oxford, UK: Oxford University Press. Quoted on p. 1.
58. See Davies, N. (1989). Sexual conflict and the polygamy threshold. *Animal Behaviour, 38*(2), 226–234. Thanks to Mark Elgar for alerting me to this example.
59. Davies (1992), ibid. Quoted on p. 1.
60. Ah-King & Ahnesjö (2013), ibid. Quoted on p. 467, references removed.
61. Itani, J. (1959). Paternal care in the wild Japanese monkey, *Macaca fuscata fuscata*. *Primates, 2*(1), 61–93.

CHAPTER 2: ONE HUNDRED BABIES?

1. Einon, D. (1998). How many children can one man have? *Evolution and Human Behavior, 19*(6), 413–426. Quoted on p. 414. All subsequent references to Einon refer to this citation.
2. Bullough, V. L. (2001). Introduction. In V. L. Bullough, B. Appleby, G. Brewer, C. M. Hajo, & E. Katz (Eds.), *Encyclopedia of birth control* (pp. xi–xv). Santa Barbara, CA: ABC-CLIO.
3. Bullough (2001), ibid.
4. Schmitt, D. P. (2003). Universal sex differences in the desire for sexual variety: Tests from 52 nations, 6 continents, and 13 islands. *Journal of Personality and Social Psychology, 85*(1), 85–104. Quoted on p. 87, references removed.
5. Wilcox, A. J., Dunson, D. B., Weinberg, C. R., Trussell, J., & Baird, D. D. (2001). Likelihood of conception with a single act of intercourse: Providing benchmark rates for assessment of postcoital contraceptives. *Contraception, 63*(4), 211–215. See Figure 1 on p. 212, and accompanying text.
6. This assumes that each coital act was independent—that is, that having sex with Woman A on Tuesday doesn't affect the probability of conception with Woman B on Wednesday. See also Tuana, N. (2004). Coming to understand: Orgasm and the epistemology of ignorance. *Hypatia, 19*(1), 194–232.
7. In a later article, Schmitt acknowledges that sex with 100 women would "rarely, if ever" result in 100 offspring. However, he suggests that the odds would be increased by "repeated matings" with the same woman in the fertile period. The plausibility of achieving this feat of carefully timed sexual conquests 100 times is discussed later in the main text. Schmitt, D. P. (2005). Sociosexuality from Argentina to Zimbabwe: A 48-nation study of sex, culture, and strategies of human mating. *Behavioral and Brain Sciences, 28*(2), 247–275. Quoted on p. 249.
8. The probability of each of the one hundred women having one child (from one coital act), where the probability of clinical pregnancy per coital act is 3.1 per cent and the probability of a live birth from a clinical pregnancy

is 90 per cent, assuming independence. Based on data from Wilcox et al. (2001), ibid.

9. For review of such data, see Haselton, M. G., & Gildersleeve, K. (2011). Can men detect ovulation? *Current Directions in Psychological Science, 20*(2), 87–92.
10. Brewis, A., & Meyer, M. (2005). Demographic evidence that human ovulation is undetectable (at least in pair bonds). *Current Anthropology, 46*(3), 465–471. Women using chemical contraceptives were excluded from the analysis, and male control of sexual behaviour didn't affect the results.
11. Laden, G. (September 9, 2011). Coming to terms with the female orgasm. Retrieved from http://scienceblogs.com/gregladen/2011/09/09/coming-to-terms-with-the-femal/ on January 23, 2015.
12. Schmitt later makes the point that a man who has sex with one hundred women will have greater reproductive success than a woman who has sex with one hundred men. This seems reasonable, but the logistics of achieving this given the two- to three-day time frame for identifying and courting the next available woman in the fertile phase of her menstrual cycle seems implausible for most men, to say the least. Schmitt (2005), ibid.
13. The probability of each of the one hundred women having one child (from one coital act) where the probability of a clinical pregnancy per coital act is 8.6 per cent and the probability of a live birth from a clinical pregnancy is 90 per cent, assuming independence. Based on data from Wilcox et al. (2001), ibid, Table 1, p. 213.
14. If you're reading this book, this hasn't happened yet. Mann, A. (February 15, 2013). Odds of death by asteroid? Lower than plane crash, higher than lightning. *Wired*. Retrieved from http://www.wired.com/2013/02/asteroid-odds/ on December 30, 2015.
15. Betzig, L. (2012). Means, variances, and ranges in reproductive success: Comparative evidence. *Evolution and Human Behavior, 33*(4), 309–317. See Table 1, p. 310.
16. Brown, G. R., Laland, K. N., & Mulder, M. B. (2009). Bateman's principles and human sex roles. *Trends in Ecology and Evolution, 24*(6), 297–304.
17. Based on a probability of "success" (that is, birth) on each sexual encounter of 0.9 x 0.031, based on previously used data for these calculations. The probability that the number of "failures" (no baby) observed before 2 successes have occurred will be 136 or less, is 0.9. Or to put it differently, you need to be willing to observe 136 failures for the probability to have observed 2 successes to be 0.9. This is obviously a probabilistic statement—some men might have two "successes" right away—but the probability of this occurring is very low (0.0008). Many thanks to Carsten Murawski for his assistance with this calculation.
18. Drea (2005), ibid. Quoted on p. 916.
19. A point made by Einon (1998), ibid., citing an account of the lack of resource acquisition and status differentials among the !Kung-San, described by Broude, G. J. (1993). Attractive single gatherer wishes to meet rich, powerful hunter for good time under mongongo tree. *Behavioral and Brain Sciences,*

16(2), 287–289. That hierarchies of wealth are not observed in African foraging societies (like the !Kung San or the Hadza) is also noted by Hrdy, S. B. (2000). The optimal number of fathers: Evolution, demography, and history in the shaping of female mate preferences. *Annals of the New York Academy of Sciences, 907*(1), 75–96.

20. Fuentes, A. (2005). Ethnography, cultural context, and assessments of reproductive success matter when discussing human mating strategies. *Behavioral and Brain Sciences, 28*(2), 284–285. Quoted on p. 285.
21. For example, Buss, D. M. & Schmitt, D. P. (1993). Sexual strategies theory: An evolutionary perspective on human mating. *Psychological Review, 100*(2), 204–232.
22. Schmitt (2005), ibid. Quoted on p. 249, reference removed.
23. Smiler, A. (2013). *Challenging Casanova: Beyond the stereotype of the promiscuous young male.* San Francisco, CA: Jossey-Bass. Quoted on p. 1.
24. See also analyses by William C. Pedersen and colleagues. Pedersen, W. C., Miller, L. C., Putcha-Bhagavatula, A. D., & Yang, Y. (2002). Evolved sex differences in the number of partners desired? The long and the short of it. *Psychological Science, 13*(2), 157–161; Pedersen, W. C., Putcha-Bhagavatula, A., & Miller, L. C. (2011). Are men and women really that different? Examining some of sexual strategies theory (SST)'s key assumptions about sex-distinct mating mechanisms. *Sex Roles, 64*, 629–643.
25. Alexander, M., & Fisher, T. (2003). Truth and consequences: Using the bogus pipeline to examine sex differences in self-reported sexuality. *Journal of Sex Research, 40*(1), 27–35.
26. Wiederman, M. (1997). The truth must be in here somewhere: Examining the gender discrepancy in self-reported lifetime number of sex partners. *Journal of Sex Research, 34*(4), 375–386. Quoted on p. 375. Removing respondents who have participated in paid sex only somewhat reduces the discrepancy between men's and women's reports.
27. For instance, Schmitt (2005), ibid., found less "restrained" sociosexuality in males across 48 nations (the sociosexuality measure being an amalgam of behaviours, attitudes, and desires with respect to casual versus committed sex), and greater male interest in having more than one sexual partner over every time period inquired about, from one month to thirty years, and an overall effect size for sex differences in sociosexuality of $d=0.74$ (ranging from a low of $d=0.3$ in Latvia to a high of $d=1.24$ in Morocco and the Ukraine). Similarly, Lippa found an effect size of $d=0.74$ with a briefer measure of sociosexuality in a large-scale BBC Internet survey. Lippa, R. A. (2009). Sex differences in sex drive, sociosexuality, and height across 53 nations: Testing evolutionary and social structural theories. *Archives of Sexual Behavior, 38*(5), 631–651. However, as Eagly and Wood (2005) and Ryan and Jethá (2005) note, Schmitt's study didn't include any samples from nonindustrial societies, some of which have more egalitarian gender relations than are seen in any modern, industrialized societies. Eagly, A. H., & Wood, W. (2005). Universal sex differences across patriarchal cultures ≠

evolved psychological dispositions. *Behavioral and Brain Sciences, 28*(2), 281–283; Ryan, C., & Jethá, C. (2005). Universal human traits: The holy grail of evolutionary psychology. *Behavioral and Brain Sciences*, 28(2), 292–293. A similar limitation applies to Lippa (2009).

28. The reason I am drawing on these data rather than, for instance, those of Schmitt (2003, 2005), ibid., is because of the benefit of looking at data drawn from probability sampling, rather than predominantly young college students who are neither representative of the entire population from which they are drawn, nor can be assumed to be able to accurately predict what kind of sexual relationships they will want in decades of their life yet to come. These limitations are discussed, for example, by Fuentes (2005), ibid.; Asendorpf, J. B., & Penke, L. (2005). A mature evolutionary psychology demands careful conclusions about sex differences. *Behavioral and Brain Sciences*, 28(2), 275–276.
29. See reference Table 3.1. Available at http://www.natsal.ac.uk/natsals-12/results-archived-data.aspx.
30. Ibid. See Tables 3.25, 3.17, and 3.9.
31. Ibid. See Table 3.17.
32. Ibid. See Table 8.1.
33. Ibid. See Table 8.2.
34. Data reviewed in Petersen, J. L., & Hyde, J. S. (2010). A meta-analytic review of research on gender differences in sexuality, 1993–2007. *Psychological Bulletin, 136*(1), 21–38.
35. See Table 8.2. Percentages for divorced, separated, or widowed women are 81 and 88 per cent, respectively.
36. Ibid. See Table 8.4.
37. Clark, R. D., & Hatfield, E. (1989). Gender differences in receptivity to sexual offers. *Journal of Psychology and Human Sexuality*, 2(1), 39–55. Instructions quoted from p. 49.
38. Hald, G. M., & Høgh-Olesen, H. (2010). Receptivity to sexual invitations from strangers of the opposite gender. *Evolution and Human Behavior, 31*(6), 453–458; Guéguen, N. (2011). Effects of solicitor sex and attractiveness on receptivity to sexual offers: A field study. *Archives of Sexual Behavior, 40*(5), 915–919.
39. Interestingly, a slightly different picture of female interest in casual sex emerged from a magazine investigation in which a German journalist (described as being of "observably above-average attractiveness") approached a hundred different women to ask if they would have sex with him. Not only did six of the women agree, but their willingness "was actually verified, as the requestor subsequently had sex with the women who complied." See account in Voracek, M., Hofhansl, A., & Fisher, M. L. (2005). Clark and Hatfield's evidence of women's low receptivity to male strangers' sexual offers revisited. *Psychological Reports, 97*(1), 11–20. Quoted on p. 16.
40. Tappé, M., Bensman, L., Hayashi, K., & Hatfield, E. (2013). Gender differences

in receptivity to sexual offers: A new research prototype. *Interpersona: An International Journal on Personal Relationships, 7*(2), 323–344. On a scale of 1 to 10 where 1 was labelled "No, never" and 10 was labelled "Yes, definitely," mean scores for the apartment request were 4.30 and for sex were 3.52.

41. Tappé et al. (2013), ibid. Quoted on p. 337.
42. For example, Tang-Martínez, Z. (1997). The curious courtship of sociobiology and feminism: A case of irreconcilable differences. In P. A. Gowaty (Ed.), *Feminism and evolutionary biology: Boundaries, intersections and frontiers* (pp. 116–150). Dordrecht, Netherlands: Springer.
43. Tappé et al. (2013), ibid.
44. Burt, M. R., & Estep, R. E. (1981). Apprehension and fear: Learning a sense of sexual vulnerability. *Sex Roles, 7*(5), 511–522.
45. Crawford, M., & Popp, D. (2003). Sexual double standards: A review and methodological critique of two decades of research. *Journal of Sex Research, 40*(1), 13–26.
46. Bordini, G. S., & Sperb, T. M. (2013). Sexual double standard: A review of the literature between 2001 and 2010. *Sexuality and Culture, 17*(4), 686–704.
47. Crawford & Popp (2003), ibid., referring to the work of Moffat, M. (1989). *Coming of age in New Jersey.* New Brunswick, NJ: Rutgers University Press. Quoted on pp. 19–20. Although this study is now rather dated, a more recent qualitative study of young Australians, by Michael Flood, showed that despite a slight shift in sexual attitudes, with some concerns expressed by young men about being labelled a "male slut," the meaning of this term did "not have the same moral and disciplinary weight of the term 'slut' when applied to women." Flood concludes that the "sexual double standard remains a persistent feature of contemporary heterosexual sexual and intimate relations." Flood, M. (2013). Male and female sluts: Shifts and stabilities in the regulation of sexual relations among young heterosexual men. *Australian Feminist Studies, 28*(75), 95–107. Quoted on p. 105.
48. Crawford & Popp (2003), ibid., quoting Moffat (1989), ibid., on p. 20.
49. O'Toole, E. (2015). *Girls will be girls: Dressing up, playing parts and daring to act differently.* London: Orion. Quoted on pp. 10–11.
50. From Sutton, L. A. (1995). Bitches and skankly hobags: The place of women in contemporary slang. In K. Hall & M. Bucholtz (Eds.), *Gender articulated: Language and the socially constructed self* (pp. 279–296). Cited in Crawford & Popp (2003), ibid.
51. Crawford & Popp (2003), ibid.
52. Rudman, L., Fetterolf, J. C., & Sanchez, D. T. (2013). What motivates the sexual double standard? More support for male versus female control theory. *Personality and Social Psychology Bulletin, 39*(2), 250–263.
53. Armstrong, E. A., England, P., & Fogarty, A. C. (2012). Accounting for women's orgasm and sexual enjoyment in college hookups and relationships. *American Sociological Review, 77*(3), 435–462.
54. Armstrong et al. (2012), ibid. Quoted on p. 456.
55. Conley, T. D. (2011). Perceived proposer personality characteristics and

gender differences in acceptance of casual sex offers. *Journal of Personality and Social Psychology, 100*(2), 309–329.

56. Conley, T. D., Ziegler, A., & Moors, A. C. (2013). Backlash from the bedroom: Stigma mediates gender differences in acceptance of casual sex offers. *Psychology of Women Quarterly, 37*(3), 392–407.
57. Conley (2011), ibid.
58. Conley (2011), ibid.
59. Fenigstein, A., & Preston, M. (2007). The desired number of sexual partners as a function of gender, sexual risks, and the meaning of "ideal." *Journal of Sex Research, 44*(1), 89–95. There was a similar absence of a differential contribution for health risks, although these didn't emerge as being very important in Conley's research.
60. Baumeister, R. F. (2013). Gender differences in motivation shape social interaction patterns, sexual relationships, social inequality, and cultural history. In M. K. Ryan & N. Branscombe (Eds.), *The Sage handbook of gender and psychology* (pp. 270–285). London: Sage. Quoted on p. 272.
61. Hald & Høgh-Olesen (2010), ibid. Quoted on p. 457.
62. See discussion in Fuentes, A. (2012). *Race, monogamy, and other lies they told you: Busting myths about human nature.* Berkeley: University of California Press.
63. Zuk, M. (2013). *Paleofantasy: What evolution really tells us about sex, diet, and how we live.* New York: Norton. Quoted on p. 181.
64. Starkweather, K., & Hames, R. (2012). A survey of non-classical polyandry. *Human Nature, 23*(2), 149–172. Quoted on p. 167. The factors associated with polyandry are an operational sex ratio in favour of males and, to a lesser degree, adult male mortality and male absenteeism.
65. Clarkin, P. F. (July 5, 2011). Part 1. Humans are (blank)-ogamous. *Kevishere.com.* Retrieved from http://kevishere.com/2011/07/05/part-1-humans-are-blank-ogamous/ on August 20, 2015.

CHAPTER 3: A NEW POSITION ON SEX

1. Saad, G., & Gill, T. (2003). An evolutionary psychology perspective on gift giving among young adults. *Psychology and Marketing, 20*(9), 765–784. Quoted on p. 769.
2. A number of feminist biologists have noted this tendency, and identified why it is problematic. For example, Fausto-Sterling, A. (1992). *Myths of gender: Biological theories about women and men.* New York: Basic Books; Tang-Martínez, Z. (1997). The curious courtship of sociobiology and feminism: A case of irreconcilable differences. In P. A. Gowaty (Ed.), *Feminism and evolutionary biology: Boundaries, intersections and frontiers* (pp. 116–150). Dordrecht, Netherlands: Springer; Zuk, M. (2002). *Sexual selections: What we can and can't learn about sex from animals.* Berkeley: University of California Press.
3. In these cases, Tang-Martínez (1997), ibid., notes: studying the trait in one species can offer no insights into the genetic or evolutionary origins of the trait in the other species.

4. Klein, J. G., Lowery, T. M., & Otnes, C. C. (2015). Identity-based motivations and anticipated reckoning: Contributions to gift-giving theory from an identity-stripping context. *Journal of Consumer Psychology, 25*(3), 431–448.
5. Schwartz, B. (1967). The social psychology of the gift. *American Journal of Sociology, 73*(1), 1–11. Quoted on p. 2.
6. Marks, J. (2012). The biological myth of human evolution. *Contemporary Social Science, 7*(2), 139–157. Quoted on p. 148, reference removed.
7. Marks, J. (November 10, 2013). Nulture. *Popanth*. Retrieved from http://popanth.com/article/nulture/on July 21, 2014.
8. Downey, G. (January 10, 2012). The long, slow sexual revolution (part 1) with NSFW video. *PLOS Blogs*. Retrieved from http://blogs.plos.org/neuroanthropology/2012/01/10/the-long-slow-sexual-revolution-part-1-with-nsfw-video/ on January 23, 2015. Downey suggests (in removed portion of quotation) that this is in fact the case for most sexual species.
9. See, for example, Wallen, K., & Zehr, J. L. (2004). Hormones and history: The evolution and development of primate female sexuality. *Journal of Sex Research, 41*(1), 101–112.
10. Gould, S. J., & Vrba, E. S. (1982). Exaptation: A missing term in the science of form. *Paleobiology, 8*(1), 4–15.
11. Dupré, J. (2001). *Human nature and the limits of science*. Oxford, UK: Oxford University Press. Quoted on p. 58.
12. Abramson, P., & Pinkerton, S. (2002). *With pleasure: Thoughts on the nature of human sexuality*. Oxford, UK: Oxford University Press. Quoted on p. 5.
13. Meston, C. M., & Buss, D. M. (2007). Why humans have sex. *Archives of Sexual Behavior, 36*(4), 477–507.
14. For comparisons with other primates, see Wrangham, R., Jones, J., Laden, G., Pilbeam, D., & Conklin-Brittain, N.-L. (1999). The raw and the stolen: Cooking and the ecology of human origins. *Current Anthropology, 40*(5), 567–594. They note that even in natural-fertility human populations, "the number of mating days between births is exceptionally high." Quoted on p. 573.
15. Laden, G. (September 9, 2011). Coming to terms with the female orgasm. *Science Blogs*. Retrieved from http://scienceblogs.com/gregladen/2011/09/09/coming-to-terms-with-the-femal/ on January 23, 2015.
16. Wolf, N. (2012). *Vagina: A new biography*. New York: HarperCollins. Quoted on p. 327.
17. Meston & Buss (2007), ibid. See Table 10 on p. 497.
18. Smiler, A. (2013). *Challenging Casanova: Beyond the stereotype of the promiscuous young male*. San Francisco, CA: Jossey-Bass. Quoted on p. 4.
19. The third NATSAL survey of men ages 16–74 indicated that 11 per cent of UK men have paid for sex in their lifetime, and 3.6 per cent in the past five years. (The comparable figure for women ages 16–44 was 0.1 per cent.) Jones, K. G., Johnson, A. M., Wellings, K., Sonnenberg, P., Field, N., Tanton, C., et al. (2015). The prevalence of, and factors associated with, paying for sex among men resident in Britain: Findings from the third National Survey of

Sexual Attitudes and Lifestyles (NATSAL-3). *Sexually Transmitted Infections, 91*(2), 116–123.

20. Sanders, T. (2008). Male sexual scripts: Intimacy, sexuality and pleasure in the purchase of commercial sex. *Sociology, 42*(3), 400–417. Quoted on p. 403.
21. Sanders (2008), ibid. Quoted on p. 400.
22. Holzmann, H., & Pines, S. (1982). Buying sex: The phenomenology of being a john. *Deviant Behavior, 4*(1), 89–116. Quotations from pp. 108 and 110, respectively. In about half the sample, paying for sex was reported to be motivated by a desire for companionship as well as sexual pleasure.
23. Sanders (2008), ibid. Quoted on p. 407.
24. Laden (2011), ibid. He is referring here specifically to females, but previously argues the same point for males.
25. Geary (2010), for instance, writes that "The bottom line is that the preferred mates and attendant cognitions and behaviors . . . of both sexes evolved to focus on and exploit the reproductive potential and reproductive investment of the opposite sex." Geary, D. C. (2010). *Male, female: The evolution of human sex differences* (2nd ed.). Washington, DC: American Psychological Association. Quoted on p. 211.
26. For example, Gangestad, S. W., Thornhill, R., & Garver, C. E. (2002). Changes in women's sexual interests and their partner's mate-retention tactics across the menstrual cycle: Evidence for shifting conflicts of interest. *Proceedings of the Royal Society London B, 269*, 975–982; Little, A. C., Jones, B. C., & DeBruine, L. M. (2008). Preferences for variation in masculinity in real male faces change across the menstrual cycle: Women prefer more masculine faces when they are more fertile. *Personality and Individual Differences, 45*(6), 478–482.
27. Hrdy, S. B. (2000). The optimal number of fathers: Evolution, demography, and history in the shaping of female mate preferences. *Annals of the New York Academy of Sciences, 907*(1), 75–96. Quoted on p. 90. Hrdy cites Crow, J. F. (1999). The odds of losing at genetic roulette. *Nature, 397*(6717), 293–294.
28. For a recent review, noting increased rate of de novo mutations in the sperm of older males, and their contribution to genetic disease, see Veltman, J. A., & Brunner, H. G. (2012). De novo mutations in human genetic disease. *Nature Reviews Genetics, 13*(8), 565–575.
29. Pillsworth, E. G. (2008). Mate preferences among the Shuar of Ecuador: Trait rankings and peer evaluations. *Evolution and Human Behavior, 29*(4), 256–267. Quoted on p. 257.
30. Marlowe, F. W. (2004). Mate preferences among Hadza hunter-gatherers. *Human Nature, 15*(4), 365–376. A constructed variable that together combined looks, age, and fertility (which in particular was cited as being more important to males) yielded a higher score of importance to males than to females.
31. Pillsworth (2008), ibid.
32. Hrdy, S. B. (1997). Raising Darwin's consciousness. *Human Nature, 8*(1), 1–49. Quoted on p. 4.

33. See, for example, Buss, D. M. (1989). Sex differences in human mate preferences: Evolutionary hypotheses tested in 37 cultures. *Behavioral and Brain Sciences, 12*(1), 1–14.
34. Dupré (2001), ibid. Quoted on p. 51.
35. For example, Gwynne, D. T., & Simmons, L. W. (1990). Experimental reversal of courtship roles in an insect. *Nature, 346*(6280), 172–174.
36. Zentner, M., & Mitura, K. (2012). Stepping out of the caveman's shadow: Nations' gender gap predicts degree of sex differentiation in mate preferences. *Psychological Science, 23*(10), 1176–1185.
37. Wood, W., & Eagly, A. H. (2013). Biology or culture alone cannot account for human sex differences and similarities. *Psychological Inquiry, 24*(3), 241–247.
38. Wood & Eagly (2013), ibid. Quoted on p. 245, references removed. See Sweeney, M. M. (2002). Two decades of family change: The shifting economic foundations of marriage. *American Sociological Review, 67*(1), 132–147; Sweeney, M. M., & Cancian, M. (2004). The changing importance of white women's economic prospects for assortative mating. *Journal of Marriage and Family, 66*(4), 1015–1028.
39. Buston, P. M., & Emlen, S. T. (2003). Cognitive processes underlying human mate choice: The relationship between self-perception and mate preference in Western society. *Proceedings of the National Academy of Sciences, 100*(15), 8805–8810.
40. For example, the amount of variation explained for preference for wealth and status and physical appearance, by self-perception of those same attributes, was 23 per cent and 19 per cent respectively for women, and 19 per cent and 11 per cent for men. By contrast, the amount of variation for preference for wealth and status by self-perceived physical appearance was 6 per cent in females and 5 per cent in males. For the amount of variation in preference for physical attractiveness explained by wealth and status, this was 7 per cent for males and 5 per cent for females. In other words, while there was weak support for the "potentials attract" hypothesis, similarly sized, small correlations were also seen in the "wrong" sex.
41. Buston & Emlen (2003), ibid. Quoted on p. 8809.
42. Todd, P. M., Penke, L., Fasolo, B., & Lenton, A. P. (2007). Different cognitive processes underlie human mate choices and mate preferences. *Proceedings of the National Academy of Sciences, 104*(38), 15011–15016. However, these data didn't strongly support "potentials attract" either. In women, self-perceived attractiveness did correlate with the wealth and status and family commitment of the speed daters in whom they were interested, but also with healthiness and a composite measure of attractiveness and healthiness. In men, perceived wealth and status was unrelated to the self-perceived attractiveness or observer-rated attractiveness of the women they chose. However, self-perceived attractiveness was related to the attractiveness of their choices, consistent with the likes-attract hypothesis.

43. Kurzban, R., & Weeden, J. (2005). HurryDate: Mate preferences in action. *Evolution and Human Behavior, 26*(3), 227–244.
44. He, Q.-Q., Zhang, Z., Zhang, J.-X., Wang, Z.-G., Tu, Y., Ji, T., et al. (2013). Potentials-attract or likes-attract in human mate choice in China. *PLoS One, 8*(4), e59457. Quoted on p. 7.
45. Dupré (2001), ibid. Quoted on p. 68.
46. Fine, A. (1990). *Taking the devil's advice.* London: Viking. Quoted on p. 153.
47. Modenese, S. L., Logemann, B. K., & Snowdon, C. T. (2016). *What do women (and men) want?* Unpublished manuscript.
48. Downey (2012), ibid., emphasis removed. This tendency is sometimes explicit in Evolutionary Psychology texts that, for example, ask whether we are all "naturally restricted"; or "naturally unrestricted"; "Are women designed to be more promiscuous than men?"; or "Are men naturally more promiscuous than women?"; Do sex roles reverse or suppress women's "innate" sexual tendencies? All quotations from p. 265 of Schmitt, D. P. (2005). Sociosexuality from Argentina to Zimbabwe: A 48-nation study of sex, culture, and strategies of human mating. *Behavioral and Brain Sciences, 28*(02), 247–275.
49. Cook, H. (2004). *The long sexual revolution: English women, sex, and contraception 1800–1975.* Oxford, UK: Oxford University Press. Portions of this account of Cook's work and material that follows previously appeared in Fine, C. (November, 2012). The vagina dialogues: Do women really want more sex than men? *The Monthly.*
50. Cook (2004), ibid. Quoted on p. 12.
51. Cook (2004), ibid. Quoted on p. 161.
52. Cook (2004), ibid. Quoted on p. 106.
53. Fielding, M. (1928). *Parenthood: Design or accident?* London: Labour. Quoted in Cook (2004), ibid., on p. 133.
54. Reviewed in Sanchez, D. T., Fetterolf, J. C., & Rudman, L. A. (2012). Eroticizing inequality in the United States: The consequences and determinants of traditional gender role adherence in intimate relationships. *Journal of Sex Research, 49*(2–3), 168–183.
55. Schick, V. R., Zucker, A. N., & Bay-Cheng, L. Y. (2008). Safer, better sex through feminism: The role of feminist ideology in women's sexual well-being. *Psychology of Women Quarterly, 32*(3), 225–232; Yoder, J., Perry, R., & Saal, E. (2007). What good is a feminist identity? Women's feminist identification and role expectations for intimate and sexual relationships. *Sex Roles, 57*(5–6), 365–372. See also Sanchez et al. (2012), ibid.
56. Rudman, L., & Phelan, J. (2007). The interpersonal power of feminism: Is feminism good for romantic relationships? *Sex Roles*, 57(11–12), 787–799.
57. Tavris, C. (1992). *The mismeasure of woman: Why women are not the better sex, the inferior sex, or the opposite sex.* New York: Touchstone. Quoted on pp. 211–212 and p. 212, respectively.
58. Stewart-Williams, S., & Thomas, A. G. (2013). The ape that thought it was a peacock: Does evolutionary psychology exaggerate human sex differences? *Psychological Inquiry, 24*(3), 137–168. Quoted on p. 156.

CHAPTER 4: WHY CAN'T A WOMAN BE MORE LIKE A MAN?

1. Terman, L. M., & Miles, C. C. (1936). *Sex and personality*. New York: McGraw-Hill. Quoted on p. 1.
2. Rapaille, C., & Roemer, A. (2015). *Move up: Why some cultures advance while others don't*. London: Allen Lane. Quoted on p. 44.
3. Wolpert, L. (2014). *Why can't a woman be more like a man? The evolution of sex and gender*. London: Faber & Faber.
4. Casey, P. (September 30, 2014). So why can't a woman be rather more like a man? *Irish Independent*. Retrieved from http://www.independent.ie/life/health-wellbeing/so-why-cant-a-woman-be-rather-more-like-a-man-30621028.html on September 6, 2015.
5. Casey (2014), ibid.
6. Wolpert (2014), ibid. Quoted on p. 21.
7. Richardson, S. S. (2013). *Sex itself: The search for male and female in the human genome*. Chicago: University of Chicago Press. Quoted on p. 9.
8. Blackless, M., Charuvastra, A., Derryck, A., Fausto-Sterling, A., Lauzanne, K., & Lee, E. (2000). How sexually dimorphic are we? Review and synthesis. *American Journal of Human Biology, 12*(2), 151–166.
9. Joel, D. (2012). Genetic-gonadal-genitals sex (3G-sex) and the misconception of brain and gender, or, why 3G-males and 3G-females have intersex brain and intersex gender. *Biology of Sex Differences, 3*(27).
10. From http://www.isna.org/faq/y_chromosome.
11. See http://isna/org/faq/conditions/cah.
12. Fausto-Sterling, A. (1993). The five sexes: Why male and female are not enough. *Sciences, 33*(2), 20–24.
13. Fausto-Sterling, A. (1989). Life in the XY corral. *Women's Studies International Forum, 12*(3), 319–331. Quoted on p. 329. See also historical account in Richardson (2013), ibid.
14. See Richardson (2013), ibid.
15. Ainsworth, C. (2015). Sex redefined. *Nature, 518*(19 February), 288–291. Quoted on p. 289.
16. Liben, L. (2015). Probability values and human values in evaluating single-sex education. *Sex Roles, 72*(9–10), 401–426. Quoted on p. 410.
17. Or its equivalent, in species with a different chromosomal sex-determination arrangement.
18. Eric Vilain, a clinical geneticist at University of California, Los Angeles, quoted in Ainsworth (2015), ibid., on p. 289.
19. Some material in this chapter describing the work of Joel et al. (2015) in the *Proceedings of the National Academy of Sciences* was previously published in Joel, D., & Fine, C. (December 1, 2015). It's time to celebrate the fact that there are many ways to be male and female. *The Guardian*. Retrieved from http://www.theguardian.com/science/2015/dec/01/brain-sex-many-ways-to-be-male-and-female?CMP=share_btn_tw on December 3, 2015.
20. Joel, D., & Yankelevitch-Yahav, R. (2014). Reconceptualizing sex, brain and

psychopathology: Interaction, interaction, interaction. *British Journal of Pharmacology, 171*(20), 4620–4635. Quoted on p. 4621.

21. McCarthy, M., & Arnold, A. (2011). Reframing sexual differentiation of the brain. *Nature Neuroscience, 14*(6), 677–683. Quoted on p. 677.
22. For example, Alexander, G. (2003). An evolutionary perspective of sex-typed toy preferences: Pink, blue, and the brain. *Archives of Sexual Behavior, 32*(1), 7–14; Bressler, E. R., Martin, R. A., & Balshine, S. (2006). Production and appreciation of humor as sexually selected traits. *Evolution and Human Behavior, 27*(2), 121–130.
23. Joel, D. (2011). Male or female? Brains are intersex. *Frontiers in Integrative Neuroscience, 5*(57). See also McCarthy, M. M., Pickett, L. A., VanRyzin, J. W., & Kight, K. E. (2015). Surprising origins of sex differences in the brain. *Hormones and Behavior, 76*, 3–10.
24. See Joel (2011, 2012), ibid.
25. Presenting the findings of Shors, T. J., Chua, C., & Falduto, J. (2001). Sex differences and opposite effects of stress on dendritic spine density in the male versus female hippocampus. *Journal of Neuroscience, 21*(16), 6292–6297.
26. This observation is indebted to Joel.
27. McCarthy et al. (2015), ibid. Quoted on p. 6.
28. Joel, D., Berman, Z., Tavor, I., Wexler, N., Gaber, O., Stein, Y., et al. (2015). Sex beyond the genitalia: The human brain mosaic. *Proceedings of the National Academy of Sciences, 112*(50), 15468–15473. Quoted on p. 15468.
29. Joel et al. (2015), ibid.
30. The numbers of people with consistently "intermediate" brain features (neither "male-end" or "female-end") were similarly modest, never exceeding 3.6 per cent. See Table 1, p. 15469.
31. Critiques of Joel et al.'s (2015) conclusions demonstrate that statistical techniques can be used to classify brains as belonging to females or males with reasonably high accuracy. However, as Joel and colleagues point out in a response, there is no biologically meaningful sense in which brains that are close in this statistical space are more similar than those that are more distant. Moreover, statistical techniques that successfully classify sex in one data set are unsuccessful in doing so in others. See Del Guidice, M., Lippa, R. A., Puts, D. A., Bailey, D. H., Bailey, J. M., & Schmitt D. P. (2016). Joel et al.'s method systematically fails to detect large, consistent sex differences. *Proceedings of the National Academy of Sciences, 113*(14), E1965; Joel, D., Persico, A., Hänggi, J., & Berman, Z. (2016). Response to Del Guidice et al., Chekroud et al., and Rosenblatt: Do brains of females and males belong to two distinct populations? *Proceedings of the National Academy of Sciences, 113*(14), E1969–E1970.
32. Joel et al. (2015), ibid. Quoted on p. 15468.
33. Joel (2011), ibid.
34. de Vries, G., & Forger, N. (2015). Sex differences in the brain: A whole body perspective. *Biology of Sex Differences, 6*(1), 1–15. Quoted on p. 2.
35. de Vries, G. J., & Södersten, P. (2009). Sex differences in the brain: The

relation between structure and function. *Hormones and Behavior, 55*(5), 589–596. Quoted on p. 594, references removed.

36. For numerous examples, see Fine, C. (2010a). *Delusions of gender: How our minds, society, and neurosexism create difference.* New York: Norton. For discussion of a recent, high-profile example, see Fine, C. (December 3, 2013). New insights into gendered brain wiring, or a perfect case study in neurosexism? *The Conversation.* Retrieved from https://theconversation.com/new-insights-into-gendered-brain-wiring-or-a-perfect-case-study-in-neurosexism-21083 on December 4, 2013. For analysis of two years of the human neuroimaging scientific literature, documenting frequency of such "reverse inferences," see Fine, C. (2013a). Is there neurosexism in functional neuroimaging investigations of sex differences? *Neuroethics, 6*(2), 369–409.
37. For explicit statements to this effect, see Fine, C. (2010a, 2013a), ibid.; Fine, C. (2013b). Neurosexism in functional neuroimaging: From scanner to pseudoscience to psyche. In M. Ryan & N. Branscombe (Eds.), *The Sage handbook of gender and psychology* (pp. 45–60). Thousand Oaks, CA: Sage; Fine, C., Joel, D., Jordan-Young, R. M., Kaiser, A., & Rippon, G. (December 15, 2015). Why males ≠ Corvettes, females ≠ Volvos, and scientific criticism ≠ ideology. *Cerebrum*; Fine, C. (2014). His brain, her brain? *Science 346*(6212), 915–916.
38. For arguments to this effect, see, for example, Cahill, L. (2006). Why sex matters for neuroscience. *Nature Reviews Neuroscience, 7,* 477–484; de Vries, G., & Forger, N. (2015). Sex differences in the brain: A whole body perspective. *Biology of Sex Differences, 6*(1), 1–15; McCarthy, M., Arnold, A., Ball, G., Blaustein, J., & de Vries, G. J. (2012). Sex differences in the brain: The not so inconvenient truth. *Journal of Neuroscience, 32*(7), 2241–2247.
39. Einstein, G. (May 8, 2014). When does a difference make a difference? Examples from situated neuroscience. *Neurogenderings III.* University of Lausanne, Switzerland, May 8–10, 2014. Podcast available at http://wp.unil.ch/neurogenderings3/podcasts/.
40. Yan, H. (August 8, 2015). Donald Trump's "blood" comment about Megyn Kelly draws outrage. *CNN.* Retrieved from edition.cnn.com/2015/08/08/politics/donald-trump-cnn-megyn-kelly-comment/ on December [illegible]
41. Schwartz, D., Romans, S., Meiyappan, S., De Souza, M., & Einstein, G. (2012). The role of ovarian steriod hormones in mood. *Hormones and Behavior, 62*(4), 448–454. See also Romans, S. E., Kreindler, D., Asllani, E., Einstein, G., Laredo, S., Levitt, A., et al. (2013). Mood and the menstrual cycle. *Psychotherapy and Psychosomatics, 82*(1), 53–60.
42. Einstein (2014), ibid.
43. Moore, C. (1995). Maternal contributions to mammalian reproductive development and the divergence of males and females. *Advances in the Study of Behavior, 24,* 47–118. For discussion of this point in relation to human neuroimaging, see Fine (2010), ibid.; Hoffman, G. (2012). What, if anything, can neuroscience tell us about gender differences? In R. Bluhm, A. Jacobson, & H. Maibom (Eds.), *Neurofeminism: Issues at the intersection of feminist theory and cognitive science* (pp. 30–55). Basingstoke, UK: Palgrave Macmillan.

44. Maestripieri, D. (January 14, 2012). Gender differences in personality are larger than previously thought. *Psychology Today*. Retrieved from https://www.psychologytoday.com/blog/games-primates-play/201201/gender-differences-in-personality-are-larger-previously-thought on February 3, 2016.
45. Wade, L. (September 19, 2013). Sex shocker! Men and women aren't that different. *Salon*. Retrieved from http://www.salon.com/2013/09/18/sex_shocker_men_and_women_arent_that_different/ on September 25, 2015.
46. de Vries & Södersten (2009), ibid. Quoted on p. 594.
47. de Vries, G. (2004). Sex differences in adult and developing brains: Compensation, compensation, compensation. *Endocrinology, 145*(3), 1063–1068.
48. Fausto-Sterling, A. (2012). *Sex/gender: Biology in a social world*. New York & London: Routledge.
49. Fausto-Sterling (2012), ibid. Quoted on p. 31, references removed. Fausto-Sterling cites the work of Gahr and colleagues. Gahr, M., Metzdorf, R., Schmidl, D., & Wickler, W. (2008). Bi-directional sexual dimorphisms of the song control nucleus HVC in a songbird with unison song. *PLoS One, 3*(8), e3073; Gahr, M., Sonnenschein, E., & Wickler, W. (1998). Sex difference in the size of the neural song control regions in a dueting songbird with similar song repertoire size of males and females. *Journal of Neuroscience, 18*(3), 1124–1131.
50. See McCarthy et al. (2015), ibid.
51. See McCarthy & Arnold (2011), ibid.
52. Moore, C. (1984). Maternal contributions to the development of masculine sexual behavior in laboratory rats. *Developmental Psychobiology, 17*(4), 347–356; Moore, C., Dou, H., & Juraska, J. (1992). Maternal stimulation affects the number of motor neurons in a sexually dimorphic nucleus of the lumbar spinal cord. *Brain Research, 572*, 52–56.
53. See Auger, A. P., Jessen, H. M., & Edelmann, M. N. (2011). Epigenetic organization of brain sex differences and juvenile social play behavior. *Hormones and Behavior, 59*(3), 358–363; de Vries & Forger (2015), ibid.
54. West, M. J., & King, A. P. (1987). Settling nature and nurture into an ontogenetic niche. *Developmental Psychobiology, 20*(5), 549–562. See also Lickliter, R. (2008). The growth of developmental thought: Implications for a new evolutionary psychology. *New Ideas in Psychology, 26*(3), 353–369.
55. Griffiths, P. E. (2002). "What is innateness?" *The Monist, 85*(1), 70–85. Quoted on p. 74.
56. Chack, E. (January 25, 2014). 21 pointlessly gendered products. *BuzzFeed*. Retrieved from http://www.buzzfeed.com/erinchack/pointlessly-gendered-products#.abowqEoPQK on February 6, 2016.
57. Fausto-Sterling, A. (2016). How else can we study sex differences in early infancy? *Developmental Psychobiology, 58*(1), 5–16.
58. de Vries & Forger (2015), ibid. Quoted on p. 11.
59. See discussion in Fine, C. (2015). Neuroscience, gender, and "development to" and "from": The example of toy preferences. In J. Clausen & N. Levy

(Eds.), *Handbook of neuroethics* (pp. 1737–1755). Dordrecht, Netherlands: Springer.

60. For discussion of the contrast between "development to" and "development from" conceptions of development, see Moore, C. L. (2002). On differences and development. In D. J. Lewkowicz & R. Lickliter (Eds.), *Conceptions of development: Lessons from the laboratory* (pp. 57–76). New York: Psychology Press.
61. For example, "The functional significance of an environmental stimuli (like [simulated maternal grooming]) altering sex differences remains unclear, but this may serve to prepare the offspring for the type of environment it will encounter as an adult." Edelmann, M. N., & Auger, A. P. (2011). Epigenetic impact of simulated maternal grooming on estrogen receptor alpha within the developing amygdala. *Brain, Behavior, and Immunity, 25*(7), 1299–1304. Quoted on p. 1303.
62. Henrich, J., & McElreath, R. (2003). The evolution of cultural evolution. *Evolutionary Anthropology, 12*(3), 123–135. Quoted on p. 123.
63. See http://www.broadwayworld.com/bwwtv/tvshows/WIFE-SWAP-/about.
64. Pagel, M. (2012). *Wired for culture: Origins of the human social mind.* New York: Norton. Quoted on p. 2.
65. Wood, W., & Eagly, A. (2012). Biosocial construction of sex differences and similarities in behavior. In J. Olson & M. Zanna (Eds.), *Advances in experimental social psychology* (Vol. 46, pp. 55–123). Burlington, MA: Academic Press. Quoted on p. 56.
66. For example, Starkweather, K., & Hames, R. (2012). A survey of non-classical polyandry. *Human Nature, 23*(2), 149–172.
67. Wood & Eagly (2012), ibid. Quoted on p. 57.
68. Wood & Eagly (2012), ibid., indeed suggest that, to a considerable degree, gender roles recruit neurohormonal mechanisms.
69. See Wood & Eagly (2012), ibid.
70. Goldstein, J. (2001). *War and gender: How gender shapes the war system and vice versa.* Cambridge, UK: Cambridge University Press; van Wagtendonk, A. (August 21, 2014). Female Kurdish fighters take arms against Islamic State extremists. *PBS NewsHour The Rundown.* Retrieved from http://www.pbs.org/newshour/rundown/female-kurdish-fighters-take-arms-islamic-state-extremists/ on September 8, 2014.
71. Wood & Eagly (2012), ibid. Quoted on p. 57.
72. Hyde, J. (2005). The gender similarities hypothesis. *American Psychologist, 60*(6), 581–592.
73. These and subsequent interpretations of effect sizes are drawn from Table 1 of Coe, R. (2012). It's the effect size, stupid: What the "effect size" is and why it is important. Annual Conference of the British Educational Research Association, September 12–14, 2002, University of Exeter, Devon. Retrieved from www.leeds.ac.uk/educol/documents/00002182.htm on December 31, 2015.

74. Zell, E., Krizan, Z., & Teeter, S. R. (2015). Evaluating gender similarities and differences using metasynthesis. *American Psychologist, 70*(1), 10–20.
75. Carothers, B. J. & Reis, H. T. (2013). Men and women are from Earth: Examining the latent structure of gender. *Journal of Personality and Social Psychology, 104*(2), 385–407.
76. Reis, H. T., & Carothers, B. J. (2014). Black and white or shades of gray? Are gender differences categorical or dimensional? *Current Directions in Psychological Science, 23*(1), 19–26. Quoted on p. 23.
77. Schwartz, S. H., and Rubel, T. (2005). Sex differences in value priorities: Cross-cultural and multimethod studies. *Journal of Personality and Social Psychology, 89*(6), 1010–1028. Schwartz & Rubel assessed the importance of basic values across seventy countries. Across samples, the median effect size was d=0.15, and the largest was d=0.32 (for power). Age and culture explained considerably more variance than did sex.
78. Patten, E., & Parker, K. (2012). *A gender reversal on career aspirations.* Pew Research Center. Retrieved from www.pewsocialtrends.org/2012/04119/a-gender-reversal-on-career-aspirations/ on March 24, 2015.
79. See Chapter 5 of Fuentes, A. (2012). *Race, monogamy, and other lies they told you: Busting myths about human nature.* Berkeley: University of California Press.
80. See reported meta-analytic results in Hyde (2005), ibid.
81. Cameron, D. (2007). *The myth of Mars and Venus: Do men and women really speak different languages?* Oxford, UK: Oxford University Press. Discussing the work of Kulick, D. (1993). Speaking as a woman: Structure and gender in domestic arguments in a New Guinea village. *Cultural Anthropology, 8*(4), 510–541.
82. Archer, J., & Coyne, S. M. (2005). An integrated review of indirect, relational, and social aggression. *Personality and Social Psychology Review 9*(3), 212–230. Quoted on p. 212.
83. See, for example, the meta-analysis by Archer, J. (2004). Sex differences in aggression in real-world settings: A meta-analytic review. *Review of General Psychology, 8*(4), 291–322.
84. Su, R., Rounds, J., & Armstrong, P. I. (2009). Men and things, women and people: A meta-analysis of sex differences in interests. *Psychological Bulletin, 135*(6), 859–884.
85. Lippa, R. A., Preston, K., & Penner, J. (2014). Women's representation in 60 occupations from 1972 to 2010: More women in high-status jobs, few women in things-oriented jobs. *PLOS One, 9*(5).
86. Valian, V. (2014). Interests, gender, and science. *Perspectives on Psychological Science, 9*(2), 225–230. Quoted on p. 226.
87. Valian (2014), ibid. Quoted on p. 227. Valian refers here to a different, shorter interests inventory.
88. That the traditionally feminine job of nursing requires this kind of "systemizing" is a point made by Jordan-Young, R. (2010). *Brain storm: The flaws in the science of sex differences.* Cambridge, MA: Harvard University Press.

89. Cahill, L. (July 11, 2014). Equal ≠ the same: Sex differences in the human brain. *Cerebrum*. Cahill cites two studies in support. The first are the categorical sex differences found for strongly sex-stereotyped activities (such as playing golf and taking baths). Carothers & Reis (2013), ibid. However, as noted earlier in the text, Carothers and Reis did not find categorical sex differences on any psychological traits studied. Second, Cahill cites a study that combined fifteen different personality measures into a "global personality score," reporting that the overlap between the sexes on this multidimensional measure was much less than for a single personality measure: just 18 per cent. Del Giudice, M., Booth, T., & Irwing, P. (2012). The distance between Mars and Venus: Measuring global sex differences in personality. *PLoS One, 7*(1), e29265. One legitimate rationale of the study was to look separately at the different factors that make up the Big Five, since sex differences in submeasures that make up each overarching personality trait may to some extent cancel one another out. But Swansea University psychologists Steve Stewart-Williams and Andrew Thomas argue that an important feature of the multidimensional statistic used to create the "global personality score" is that the more dimensions you add, the bigger the statistic gets. "This has an awkward implication. . . . Even for very similar populations—New Zealanders and Australians, for example—there will inevitably be many variables for which there are small average differences. If you were to take enough of these variables and treat them as a single multidimensional variable, you could use Del Giudice's method to 'prove' that, psychologically, New Zealanders and Australians are virtually different species." Stewart-Williams, S., & Thomas, A. G. (2013). The ape that thought it was a peacock: Does Evolutionary Psychology exaggerate human sex differences? *Psychological Inquiry, 24*(3), 137–168. Quoted on p. 168. Additionally, Hyde (2014) has observed that Del Giudice et al.'s multidimensional measure doesn't relate to any known concept in personality, making it difficult to know how to interpret the result. Hyde, J. (2014). Gender similarities and differences. *Annual Review of Psychology, 65*(1), 373–398.
90. Terman & Miles (1936), ibid. See Lippa, R. A. (2002). *Gender, nature, and nurture*. Mahwah, NJ: Lawrence Earlbaum.
91. Spence, J. T., Helmreich, R. L., & Stapp, J. (1974). The Personal Attributes Questionnaire: A measure of sex role stereotypes and masculinity-femininity. *JSAS Catalog of Selected Documents in Psychology, 4*, 43–44; The Bem Sex Role Inventory, Bem, S. (1974). The measurement of psychological androgyny. *Journal of Consulting and Clinical Psychology, 42*(2), 155–162.
92. In particular, the multidimensional gender identity theory of Spence, J. T. (1993). Gender-related traits and gender ideology: Evidence for a multifactorial theory. *Journal of Personality and Social Psychology, 64*(4), 624–635. See also Egan, S. K., & Perry, D. G. (2001). Gender identity: A multidimensional analysis with implications for psychosocial adjustment. *Developmental Psychology, 37*(4), 451–463.
93. Wolpert (2014), ibid. Quoted on p. 179.

94. Valian, V. (2014). Developmental biology: Splitting the sexes. *Nature, 513*(7516), 32. Quoted on p. 32.
95. Cimpian, A., & Markman, E. M. (2011). The generic/nongeneric distinction influences how children interpret new information about social others. *Child Development, 82*(2), 471–492. Sample explanations from Table 1, p. 477.
96. Cimpian & Markman (2011), ibid. Quoted on p. 473.
97. Browne, K. R. (2011). Evolutionary Psychology and sex differences in workplace patterns. In G. Saad (Ed.), *Evolutionary psychology in the business sciences* (pp. 71–94). Heidelberg, Germany: Springer. Quoted on p. 71.
98. Carothers & Reis (2013), ibid. A similar point is made in relation to "non-kinds" in psychiatric classification by Haslam, N. (2002). Kinds of kinds: A conceptual taxonomy of psychiatric categories. *Philosophy, Psychiatry, and Psychology, 9*(3), 203–217.

CHAPTER 5: SKYDIVING WALLFLOWERS

1. Hoffman, M., & Yoeli, E. (Winter, 2013). The risks of avoiding a debate on gender differences. *Rady Business Journal*.
2. Baker M. D., Jr., & Maner, J. K. (2008). Risk-taking as a situationally sensitive male mating strategy. *Evolution and Human Behavior, 29*(6), 391–395. Quoted on p. 392, references removed. The authors propose that additional benefits of male displays of risk taking are attracting allies and scaring off competitors.
3. Baker & Maner (2008), ibid. Quoted on p. 392, reference removed.
4. Hoffman & Yoeli (2013), ibid.
5. However, it should be noted that economists regard risk taking and competition as separate concepts—the former involves engagement with an unpredictable world, while the latter involves engagement with unpredictable others.
6. Niederle, M., & Vesterlund, L. (2011). Gender and competition. *Annual Review of Economics, 3*(1), 601–630. Quoted on p. 602.
7. Adams, R. (January 21, 2015). Gender gap in university admissions rises to record level. *The Guardian*. Retrieved from http://www.theguardian.com/education/2015/jan/21/gender-gap-university-admissions-record on May 14, 2015.
8. For example, the Choice Dilemma Scale: Kogan, N., & Wallach, M. (1964). *Risk-taking: A study in cognition and personality*. New York: Holt.
9. For example, many of the studies on competition discussed later in the chapter "control" for risk-taking propensity by including a lottery task to assess risk taking. The unstated assumption is presumably that this particular measure of risk taking also captures propensity to take risk in a different domain, namely, taking the risk of competing against peers.
10. Wilson, M., & Daly, M. (1985). Competitiveness, risk taking, and violence: The young male syndrome. *Ethology and Sociobiology, 6*(1), 59–73. Quoted on p.66.

11. Johnson, J., Wilke, A., & Weber, E. U. (2004). Beyond a trait view of risk taking: A domain-specific scale measuring risk perceptions, expected benefits, and perceived-risk attitudes in German-speaking populations. *Polish Psychological Bulletin, 35*(3), 153–163. Quoted on p. 153. See also early discussion of this issue in Slovic, P. (1964). Assessment of risk taking behavior. *Psychological Bulletin, 61*(3), 220–233.
12. MacCrimmon, K., & Wehrung, D. (1985). A portfolio of risk measures. *Theory and Decision, 19*(1), 1–29.
13. Weber, E. U., Blais, A.-R., & Betz, N. E. (2002). A domain-specific risk-attitude scale: Measuring risk perceptions and risk behaviors. *Journal of Behavioral Decision Making, 15*(4), 263–290. Weber's group then went on to find the same domain specificity of risk-taking propensity in a large sample of young Germans. Johnson et al. (2004), ibid.
14. Hanoch, Y., Johnson, J. G., & Wilke, A. (2006). Domain specificity in experimental measures and participant recruitment: An application to risk-taking behavior. *Psychological Science, 17*(4), 300–304.
15. Weber et al. (2002), ibid.
16. Cooper, A. C., Woo, C. Y., & Dunkelberg, W. C. (1988). Entrepreneurs' perceived chances for success. *Journal of Business Venturing, 3*(2), 97–108.
17. Weber et al. (2002), ibid. This is in line with arguments made by many in the field that risk per se—the possibility of loss—is always "repugnant" (p. 265). See also Yates, J., & Stone, E. (1992). The risk construct. In K. Yates (Ed.), *Risk-taking behavior*. New York: Wiley.
18. Keyes, R. (1985). *Chancing it: Why we take risks*. Boston: Little, Brown. Quoted on pp. 10 and 9, respectively.
19. Yates & Stone (1992), ibid. Quoted on p. 2.
20. Keyes (1985), ibid. Quoted on p. 6.
21. Weber et al. (2002), ibid.; Johnson et al. (2004), ibid.; Harris, C. R., Jenkins, M., & Glaser, D. (2006). Gender differences in risk assessment: Why do women take fewer risks than men? *Judgment and Decision Making, 1*(1), 48–63. A similar principle was found to be at work in explaining cross-cultural differences in buying prices for risky financial options: these were found to be due to differences in perceptions of the risks of the financial options, not risk attitudes. Weber, E. U., & Hsee, C. (1998). Cross-cultural differences in risk perception, but cross-cultural similarities in attitudes towards perceived risk. *Management Science, 44*(9), 1205–1217.
22. Byrnes, J. P., Miller, D. C., & Schafer, W. D. (1999). Gender differences in risk taking: A meta-analysis. *Psychological Bulletin, 125*(3), 367–383.
23. Byrnes et al. (1993), ibid. Quoted on p. 377.
24. Nelson, J. A. (2014). The power of stereotyping and confirmation bias to overwhelm accurate assessment: The case of economics, gender, and risk aversion. *Journal of Economic Methodology, 21*(3), 211–231. Nelson uses the examples of domestic violence, and pregnancy and childbirth.
25. Mortality ratio for pregnancy in the United States in 2011: 17.8 deaths per 100,000 live births; available at http://www.cdc.gov/reproductivehealth/

MaternalInfantHealth/PMSS.html. Mortality ratio for skydiving in 2014: 0.75 deaths per 100,000 jumps; available at (http://www.uspa.org/AboutSkydiving/SkydivingSafety/tabid/526/Default.aspx).

26. See http://www.osteopathic.org/osteopathic-health/about-your-health/health-conditions-library/womens-health/Pages/high-heels.aspx.
27. One of the generous colleagues who reviewed this book commented in the margin that this joke seemed a bit too corny for me. As you can see, he was wrong.
28. Weber et al. (2002), ibid.; Johnson et al. (2004), ibid.; Harris et al. (2006), ibid.
29. Harris et al. (2006), ibid. This turned out to be because women saw the chances of success as more probable and the positive consequences as better.
30. For example, in the financial domain: Wang, M., Keller, C., & Siegrist, M. (2011). The less you know, the more you are afraid of: A survey on risk perceptions of investment products. *Journal of Behavioral Finance, 12*, 9–19; Weber, E. U., Siebenmorgen, N., & Weber, M. (2005). Communicating asset risk: How name recognition and the format of historic volatility information affect risk perception and investment decisions. *Risk Analysis, 25*(3), 597–609. In relation to health and leisure risks, see Song, H., & Schwarz, N. (2009). If it's difficult to pronounce, it must be risky: Fluency, familiarity, and risk perception. *Psychological Science, 20*(2), 135–138.
31. Sunstein, C. (1996). Social norms and social roles. *Columbia Law Review, 96*(4), 903–968. Quoted on p. 913, reference/footnote removed from first quotation.
32. This doesn't seem to be due to differences in relevant knowledge. For example, female members of the British Toxicology Society provided higher risk ratings of chemicals than did their male counterparts. Slovic, P., Malmfors, T., Mertz, C., Neil, N., & Purchase, I. F. (1997). Evaluating chemical risks: Results of a survey of the British Toxicology Society. *Human and Experimental Toxicology, 16*(6), 289–304.
33. Flynn, J., Slovic, P., & Mertz, C. K. (1994). Gender, race, and perception of environmental health risks. *Risk Analysis, 14*(6), 1101–1108.
34. See Finucane, M. L., Slovic, P., Mertz, C. K., Flynn, J., & Satterfield, T. A. (2000). Gender, race, and perceived risk: The "white male" effect. *Health, Risk and Society, 2*(2), 159–172; Kahan, D. M., Braman, D., Gastil, J., Slovic, P., & Mertz, C. K. (2007). Culture and identity-protective cognition: Explaining the white-male effect in risk perception. *Journal of Empirical Legal Studies, 4*(3), 465–505. See also Palmer, C. (2003). Risk perception: Another look at the "white male" effect. *Health, Risk and Society, 5*(1), 71–83. This survey found that the "white male effect" extended to Taiwanese American males, for health and technology risks.
35. Kahan, D. (October 7, 2012). Checking in on the "white male effect" for risk perception. *The Cultural Cognition Project at Yale Law School*. Retrieved from http://www.culturalcognition.net/blog/2012/10/7/checking-in-on-the-white-male-effect-for-risk-perception.html on November 7, 2014.

36. Olofsson, A., & Rashid, S. (2011). The white (male) effect and risk perception: Can equality make a difference? *Risk Analysis, 31*(6), 1016–1032.
37. Slovic, P., Finucane, M., Peters, E., & MacGregor, D. G. (2002). Rational actors or rational fools: Implications of the affect heuristic for behavioral economics. *Journal of Socio-Economics, 31*(4), 329–342. Quoted on p. 333. Confirmatory evidence was also found by Weber et al. (2002), ibid., who similarly found negative correlations between perceived risks and perceived benefits.
38. Flynn et al. (1994), ibid. Quoted on p. 1107.
39. Kahan (2012), ibid.
40. Flynn et al. (1994), ibid. Quoted on p. 1107.
41. Rawn, C. D., & Vohs, K. D. (2011). People use self-control to risk personal harm: An intra-interpersonal dilemma. *Personality and Social Psychology Review, 15*(3), 267–289.
42. For example, Prentice, D., & Carranza, E. (2002). What women and men should be, shouldn't be, are allowed to be, and don't have to be: The contents of prescriptive gender stereotypes. *Psychology of Women Quarterly, 26*(4), 269–281.
43. For example, Bowles, R., Babcock, L., & Lai, L. (2007). Social incentives for gender differences in the propensity to initiate negotiations: Sometimes it does hurt to ask. *Organizational Behavior and Human Decision Processes, 103*, 84–103; Rudman, L., & Phelan, J. E. (2008). Backlash effects for disconfirming gender stereotypes in organizations. *Research in Organizational Behavior, 28*, 61–79.
44. Hoffman & Yoeli (2013), ibid.
45. Small, D. A., Gelfand, M., Babcock, L., & Gettman, H. (2007). Who goes to the bargaining table? The influence of gender and framing on the initiation of negotiation. *Journal of Personality and Social Psychology, 93*(4), 600–613. Quoted on p. 610.
46. Gerhart, B., & Rynes, S. (1991). Determinants and consequences of salary negotiations by male and female MBA graduates. *Journal of Applied Psychology, 76*(2), 256–262.
47. M. Ryan (personal communication).
48. Mahalik, J. R., Locke, B. D., Ludlow, L. H., Diemer, M. A., Scott, R. P. J., Gottfried, M., et al. (2003). Development of the Conformity to Masculine Norms Inventory. *Psychology of Men and Masculinity, 4*(1), 3–25.
49. Prentice & Carranza (2002), ibid.
50. Brescoll, V. L., Dawson, E., & Uhlmann, E. L. (2010). Hard won and easily lost: The fragile status of leaders in stereotype-incongruent occupations. *Psychological Science, 21*(11), 1640–1642.
51. Frankenhuis, W. E., & Karremans, J. C. (2012). Uncommitted men match their risk taking to female preferences, while committed men do the opposite. *Journal of Experimental Social Psychology, 48*(1), 428–431. Interestingly, the information had the opposite effect on men already in a committed relationship.
52. Shan, W., Shenghua, J., Davis, H. M., Peng, K., Shao, X., Wu, Y., et al. (2012).

Mating strategies in Chinese culture: Female risk avoiding vs. male risk taking. *Evolution and Human Behavior, 33*(3), 182–192. There was also a condition in which participants thought they were observed by someone of the same sex, which had intermediate effects on behavior. Quoted on p. 183, references removed.

53. This was found to be mildly desirable in the U.S. sample and neither attractive nor unattractive in the German sample. Wilke, A., Hutchinson, J. M. C., Todd, P. M., & Kruger, D. J. (2006). Is risk taking used as a cue in mate choice? *Evolutionary Psychology, 4*, 367–393.
54. Quoted from the title of Farthing, G. W. (2007). Neither daredevils nor wimps: Attitudes toward physical risk takers as mates. *Evolutionary Psychology, 5*(4), 754–777. This study obtained similar findings for both sexes.
55. Wilke et al. (2006), ibid. Quoted on p. 388.
56. Sylwester, K., & Pawłowski, B. (2011). Daring to be darling: Attractiveness of risk takers as partners in long- and short-term sexual relationships. *Sex Roles, 64*(9–10), 695–706. Sylwester and Pawłowski (2011) compared the appeal of risk takers and avoiders in the domains of physical, financial, and social risks. Their main finding was that risk taking was more desirable in a short-term partner than in a long-term partner. A study by Bassett and Moss (2004) created different profiles of a low, medium, and high physical risk taker, and asked participants to rate the desirability of someone in romantic and non-romantic contexts. Men and women differed only in the desirability of a risk taker as a long-term partner. Bassett, J. F., & Moss, B. (2004). Men and women prefer risk takers as romantic and nonromantic partners. *Current Research in Social Psychology, 9*(10), 135–144.
57. Bassett & Moss (2004), ibid. Quoted on p. 140.
58. Wilke et al. (2004), ibid. Quoted on p. 387.
59. Apicella, C. L., Dreber, A., Gray, P. B., Hoffman, M., Little, A. C., & Campbell, B. C. (2011). Androgens and competitiveness in men. *Journal of Neuroscience, Psychology, and Economics, 4*(1), 54–62. Quoted on pp. 55–56, references removed. Note that the first two authors extensively discuss the growing recognition of the importance of female competition in a later publication. Apicella, C. L., & Dreber, A. (2015). Sex differences in competitiveness: Hunter-gatherer women and girls compete less in gender-neutral and male-centric tasks. *Adaptive Human Behavior and Physiology, 1*(3), 247–269.
60. Cashdan, E. (1998). Are men more competitive than women? *British Journal of Social Psychology, 37*(2), 213–229.
61. For example, Apicella & Dreber (2015), ibid.; Dreber et al. (2014), ibid.; Flory, J. A., Leibbrandt, A., & List, J. A. (2015). Do competitive work places deter female workers? A large-scale natural field experiment on gender differences on job-entry decisions: *Review of Economic Studies, 82*(1) 122–155; Grosse, N. D., Riener, G., & Dertwinkel-Kalt, M. (2014). *Explaining gender differences in competitiveness: Testing a theory on gender-task stereotypes*: working paper, University of Mannheim; Günther, C., Ekinci, N. A., Schwieren, C., & Strobel, M. (2010). Women can't jump?—An experiment on competitive attitudes

and stereotype threat. *Journal of Economic Behavior and Organization, 75*(3), 395–401; Wieland, A., & Sarin, R. (2012). Domain specificity of sex differences in competition. *Journal of Economic Behavior and Organization, 83*(1), 151–157.

62. See discussion in Khachatryan, K., Dreber, A., von Essen, E., & Ranehill, E. (2015). Gender and preferences at a young age: Evidence from Armenia. *Journal of Economic Behavior and Organization, 118*, 318–332. See also Sutter, M., & Glätzle-Rützler, D. (2014). Gender differences in the willingness to compete emerge early in life and persist. *Management Science, 61*(10), 2339–2354.
63. Cameron, L., Erkal, N., Gangadharan, L., & Meng, X. (2013). Little emperors: Behavioral impacts of China's one-child policy. *Science, 339*(6122), 953–957; Cárdenas, J.-C., Dreber, A., von Essen, E., & Ranehill, E. (2012). Gender differences in competitiveness and risk taking: Comparing children in Colombia and Sweden. *Journal of Economic Behavior and Organization, 83*(1), 11–23; Khachatryan et al. (2015), ibid.; Zhang, Y. (2015). *Culture, institutions, and the gender gap in competitive inclination: Evidence from the Communist experiment in China*. Available at SSRN: http://ssrn.com/abstract=2268874 or http://dx.doi.org/10.2139/ssrn.2268874.
64. Gneezy, U., Leonard, K. L., & List, J. A. (2009). Gender differences in competition: Evidence from a matrilineal and a patriarchal society. *Econometrica, 77*(5), 1637–1664.
65. Andersen, S., Ertac, S., Gneezy, U., List, J. A., & Maximiano, S. (2013). Gender, competitiveness, and socialization at a young age: Evidence from a matrilineal and a patriarchal society. *Review of Economics and Statistics, 95*(4), 1438–1443.
66. Sutter & Glätzle-Rützler (2014), ibid.
67. Kay, J. (December 10, 2013). Is it better to play it safe or to place bets that risk bankruptcy? *Financial Times*. Retrieved from http://www.ft.com/intl/cms/s/0/292e514e-60ff-11e3-b7f1-00144feabdc0.html#axzz3wbxCXczm on January 8, 2016. Kay does note that "Like all attempts to account for our behaviour by delving into our evolutionary past, this story should be taken with a large pinch of salt. But there does not have to be any historic truth in my narrative for its fundamental premise to be true. People who take foolish risks which mostly come off are likely to appear attractive mates and leaders." Given the prior evolutionary story, it seems to propose that by "people" here, he means "men."
68. Butt, C. (September 10, 2015). Female surgeons feel obliged to give sexual favours, report finds. *Sydney Morning Herald*. Retrieved from http://www.smh.com.au/national/bullying-endemic-among-surgeons-but-victims-too-scared-to-speak-up-report-finds-20150909-gjiuxl.html on September 10, 2015.
69. Silvester, M., & Perkins, M. (December 9, 2015). Shame files: Policewomen targeted for sex from the day they join the force. *The Age*. Retrieved from http://www.theage.com.au/victoria/a-new-report-says-policewomen-are-targeted-for-sex-from-the-day-they-join-20151208-glidtv.html on December 30, 2015.

CHAPTER 6: THE HORMONAL ESSENCE OF THE T-REX?

1. Adkins-Regan, E. (2005). *Hormones and animal social behavior.* Princeton, NJ: Princeton University Press. Quoted on p. 21.
2. Browne, K. R. (2012). Mind which gap? The selective concern over statistical sex disparities. *Florida International University Law Review, 8*, 271–286. Quoted on pp. 284–285, references removed.
3. For example, Hoffman, M., & Yoeli, E. (Winter, 2013). The risks of avoiding a debate on gender differences. *Rady Business Journal*; Cronqvist, H., Previtero, A., Siegel, S., & White, R. E. (2016). The fetal origins hypothesis in finance: Prenatal environment, the gender gap, and investor behavior. *Review of Financial Studies, 29*(3), 739–786.
4. Herbert, J. (2015). *Testosterone: Sex, power, and the will to win.* Oxford, UK: Oxford University Press. Quoted on p. 22.
5. Herbert, J. (May 14, 2015). Sex, cars, and the power of testosterone. *OUP Blog.* Retrieved from http://blog.oup.com/2015/05/sex-cars-testosterone/ on July 16, 2015.
6. Francis, R. C. (2004). *Why men won't ask for directions: The seductions of sociobiology.* Princeton, NJ: Princeton University Press. Quoted on p. 147. First reference to "Teststerone Rex" on p. 143.
7. For humans, the primary account is the biosocial model of Mazur (1985) and Mazur and Booth (1998). This proposes reciprocal influences between T and status seeking, in both women and men. Thus, higher levels of T promote status-seeking behaviour, and success feeds back to increase T levels (while failure decreases it). Mazur, A. (1985). A biosocial model of status in face-to-face primate groups. *Social Forces, 64*(2), 377–402; Mazur, A., & Booth, A. (1998). Testosterone and dominance in men. *Behavioral and Brain Sciences, 21*, 353–397. The other major theoretical framework is the challenge hypothesis, in which T changes are assumed to facilitate trade-offs between competition (or "challenge") and parenting. Wingfield, J. C., Hegner, R. E., Dufty, A. M., Jr., & Ball, G. F. (1990). The "challenge hypothesis": Theoretical implications for patterns of testosterone secretion, mating systems, and breeding strategies. *American Naturalist, 136*(6), 829–846.
8. van Anders, S. M. (2013). Beyond masculinity: Testosterone, gender/sex, and human social behavior in a comparative context. *Frontiers in Neuroendocrinology, 34*(3), 198–210. One primary and important purpose of this paper is to show how the common conflation of high T with "masculinity" and low T with "femininity," rather than more specifically with competition and nurturance, has led to inaccurate predictions that, for instance, high T will be associated with sexuality and low T with parenting. But, as van Anders points out, sexual behaviour can be either competitive or nurturant, or even both, and there are both nurturant and competitive elements to parenting. Van Anders shows how moving beyond the "pretheory" assumption that "T=masculinity" can help make sense of apparently contradictory findings, and guide better research.

9. What follows borrows extensively from Francis (2004), ibid.
10. Francis, R. C., Jacobson, B., Wingfield, J. C., & Fernald, R. D. (1992). Castration lowers aggression but not social dominance in male *Haplochromis burtoni* (Cichlidae). *Ethology, 90*(3), 247–255.
11. Francis, R. C., Soma, K., & Fernald, R. D. (1993). Social regulation of the brain-pituitary-gonadal axis. *Proceedings of the National Academy of Sciences, 90*(16), 7794–7798.
12. Francis et al. (1992), ibid. Quoted on p. 253.
13. Also see discussion in van Anders, S., & Watson, N. (2006). Social neuroendocrinology: Effects of social contexts and behaviors on sex steroids in humans. *Human Nature, 17*(2), 212–237.
14. Hrdy, S. B. (1986). Empathy, polyandry, and the myth of the coy female. In R. Bleier (Ed.), *Feminist approaches to science* (pp. 119–146). New York: Pergamon Press. Quoted on p. 141, referring to the work of Van den Berghe, E. (1984). *Female competition, parental care, and reproductive success in salmon.* Paper presented at Animal Behavior Society Meetings, Cheney, Washington, August 13–17.
15. Adkins-Regan (2005), ibid. Quoted on p. 51.
16. Joel, D. (2012). Genetic-gonadal-genitals sex (3G-sex) and the misconception of brain and gender, or, why 3G-males and 3G-females have intersex brain and intersex gender. *Biology of Sex Differences, 3*(27). Quoted on p. 4.
17. Sapolsky, R. (1997). *Junk food monkeys: And other essays on the biology of the human predicament*. London: Headline. Quoted on p. 127, footnote removed at the end of the second quotation.
18. See Freeman, E. R., Bloom, D. A., & McGuire, E. J. (2001). A brief history of testosterone. *Journal of Urology, 165*(2), 371–373. See also Adkins-Regan (2005), ibid.
19. Adkins-Regan (2005), ibid. Quoted on p. 3.
20. Moore, C. (1992). The role of maternal stimulation in the development of sexual behavior and its neural basis. *Annals of the New York Academy of Sciences, 662*(1), 160–177.
21. Oliveira (2004) describes this as the effect of testosterone on "somatic releasers"—that is, bodily effects that then influence the behaviour of conspecifics. Oliveira, R. F. (2004). Social modulation of androgens in vertebrates: Mechanisms and function. *Advances in the Study of Behavior, 34*, 165–239. For examples, see Table 1 on p. 172. Oliveira also provides examples of effects of testosterone on the sensory system and "effectors" (for example, effects on the muscles involved in producing mate calls in birds).
22. Account drawn from a far more detailed description provided by Adkins-Regan (2005), ibid.
23. See Adkins-Regan (2005), ibid., pp. 13–16, referring to both genomic (slower) and non-genomic (faster) effects. This is also well summarized in Oliveira, R. F. (2009). Social behavior in context: Hormonal modulation of behavioral plasticity and social competence. *Integrative and Comparative Biology, 49*(4), 423–440. See also Cardoso et al. (2015), who categorize three different types

of social plasticity: fixed alternative phenotypes (not applicable in the case of humans, but in other species); developmental plasticity (such as the transition between pre- and post-pubescence), and behavioural flexibility. To focus on the two categories relevant to humans, developmental plasticity is proposed to involve (re)organization of structures (including the brain) and epigenetic effects, while behavioural flexibility is activational and involves biochemical switching at the neural level, and transient changes in gene expression at the genomic level. Cardoso, S. D., Teles, M. C., & Oliveira, R. F. (2015). Neurogenomic mechanisms of social plasticity. *Journal of Experimental Biology, 218*(1), 140–149. See Table 1, p. 142.

24. Described in Pfaff, D. W. (2010). *Man and woman: An inside story.* Oxford, UK: Oxford University Press. See also Dufy, B., & Vincent, J. D. (1980). Effects of sex steroids on cell membrane excitability: A new concept for the action of steroids on the brain. In D. de Wied & P. van Keep (Eds.), *Hormones and the brain* (pp. 29–41). Lancaster, UK: MTP Press.
25. Adkins-Regan (2005), ibid. Quoted on p. 15.
26. See Chapter 1, Adkins-Regan (2005), ibid.
27. Adkins-Regan (2005), ibid. Quoted on p. 16.
28. Francis (2004), ibid., makes the point that "Of all the factors relevant to sexual development, steroid hormones, such as testosterone, are perhaps the easiest to measure, manipulate, and monitor, so they tend to be accorded more explanatory weight than other developmental factors." Quoted on pp. 143–144.
29. Gleason, E. D., Fuxjager, M. J., Oyegbile, T. O., & Marler, C. A. (2009). Testosterone release and social context: When it occurs and why. *Frontiers in Neuroendocrinology, 30*(4), 460–469. Quoted on p. 460.
30. See, for example, Adkins-Regan (2005), ibid., pp. 218–222. See also Adkins-Regan, E. (2012). Hormonal organization and activation: Evolutionary implications and questions. *General and Comparative Endocrinology, 176*(3), 279–285.
31. It's surprisingly hard to find male/female norms for testosterone levels. As a methods paper on T measurement by van Anders et al. (2014) makes clear, any such "norms" would have to take into account the considerable variability introduced by factors such as season, time of day, relationship status, body weight, and so on. van Anders, S. M., Goldey, K. L., & Bell, S. N. (2014). Measurement of testosterone in human sexuality research: Methodological considerations. *Archives of Sexual Behavior, 43*(2), 231–250. But with all due caveats in place, an effect size of about d=3 is estimated from samples from work by van Anders and colleagues, who commonly include both sexes in their research. For reference ranges for testosterone for children and adults, using a more sensitive measurement technique, see Kushnir, M. M., Blamires, T., Rockwood, A. L., Roberts, W. L., Yue, B., Erdogan, E., et al. (2010). Liquid chromatography: Tandem mass spectrometry assay for androstenedione, dehydroepiandrosterone, and testosterone with pediatric and adult reference intervals. *Clinical Chemistry, 56*(7), 1138–1147.
32. For example, de Vries, G. (2004). Sex differences in adult and developing

brains: Compensation, compensation, compensation. *Endocrinology, 145*(3), 1063–1068.

33. For example, Bancroft, J. (2002). Sexual effects of androgens in women: Some theoretical considerations. *Fertility and Sterility, 77*(Suppl. 4), 55–59; Bancroft, J. (2005). The endocrinology of sexual arousal. *Journal of Endocrinology, 186*(3), 411–427.
34. Sherwin, B. (1988). A comparative analysis of the role of androgen in human male and female sexual behavior: Behavioral specificity, critical thresholds, and sensitivity. *Psychobiology, 16*(4), 416–425. For evidence of sex differences in oestrogen receptors, see Gillies, G. E., & McArthur, S. (2010). Estrogen actions in the brain and the basis for differential action in men and women: A case for sex-specific medicines. *Pharmacological Reviews, 62*(2), 155–198.
35. Bancroft (2002, 2005), ibid.
36. A point made by van Anders (2013), ibid.; Adkins-Regan (2005), ibid. The absence of T research in women is also observed by Fausto-Sterling, A. (1992). *Myths of gender: Biological theories about women and men*. New York: Basic Books. It should be said, however, that the relative dearth of evidence with women was not simply due to a lack of interest in testosterone's effects on women, but was also related to technical difficulties accurately measuring their lower levels, misplaced concerns that the menstrual cycle might lead to significant variation in T (it doesn't), and the lowering of T by the use of oral contraceptives. However, these lower levels don't affect *change* in T in response to competition. For discussion of these and other issues, see van Anders et al. (2014), ibid.
37. van Anders (2013), ibid. Quoted on p. 198.
38. Healy, M., Gibney, J., Pentecost, C., Wheeler, M., & Sonksen, P. (2014). Endocrine profiles in 693 elite athletes in the postcompetition setting. *Clinical Endocrinology, 81*(2), 294–305.
39. Adkins-Regan, (2005), ibid. Quoted on p.4.
40. See discussion in Oliveira (2009), ibid.
41. Dixson, A. F., & Herbert, J. (1977). Testosterone, aggressive behavior and dominance rank in captive adult male talapoin monkeys (*Miopithecus talapoin*). *Physiology and Behavior, 19*(3), 539–543. Quoted on p. 542.
42. See Wallen, K. (2001). Sex and context: Hormones and primate sexual motivation. *Hormones and Behavior, 40*(2), 339–357.
43. Wallen (2001), ibid. Quoted on p. 340.
44. Described in Wallen (2001), ibid. Wallen notes that additional possible contributing factors are the impoverished social environment (leaving little else to do) and the absence of social repercussions, due to the lack of a full social group.
45. Wallen (2001), ibid. Quoted on p. 346.
46. Adkins-Regan (2005), ibid. Quoted on p. 3.
47. Oliveira (2009), ibid. Quoted on p. 423.
48. See, for example, previously cited articles by R. Oliveira. This is also the principle underlying Mazur's biosocial model; Mazur (1985), ibid.; Mazur

& Booth (1998), ibid. Oliveira (2009), ibid., (p. 427) summarizes it as follows: "The social interactions in which an individual participates or to which he is exposed, influence its androgen levels, which in turn will modulate perceptive, motivational, and cognitive mechanisms that will affect his subsequent behavior in social interactions." See also van Anders & Watson (2006), ibid.

49. Oliveira, R. F., Almada, V. C., & Canario, A. V. M. (1996). Social modulation of sex steroid concentrations in the urine of male cichlid fish, *Oreochromis mossambicus. Hormones and Behavior, 30*(1), 2–12.
50. Oliveira (2009), ibid. Quoted on p. 426.
51. Fuxjager, M. J., Forbes-Lorman, R. M., Coss, D. J., Auger, C. J., Auger, A. P., & Marler, C. A. (2010). Winning territorial disputes selectively enhances androgen sensitivity in neural pathways related to motivation and social aggression. *Proceedings of the National Academy of Sciences, 107*(27), 12393–12398. See also Burmeister, S. S., Kailasanath, V., & Fernald, R. D. (2007). Social dominance regulates androgen and estrogen receptor gene expression. *Hormones and Behavior, 51*(1), 164–170.
52. Oliveira (2004), ibid. Quoted on p. 194. Note that reference is not being made here to T levels, but the ratio of 11-ketotestosterone to testosterone, indicating the conversion of the latter into the former; 11-ketotestosterone is an androgen found only in teleost fish.
53. Ziegler, T. E., Schultz-Darken, N. J., Scott, J. J., Snowdon, C. T., & Ferris, C. F. (2005). Neuroendocrine response to female ovulatory odors depends upon social condition in male common marmosets, *Callithrix jacchus. Hormones and Behavior, 47*(1), 56–64. Paired non-fathers also showed T increases but the authors noted that these pairings were, at that point, short-lived.
54. Briefly but helpfully reviewed in van Anders (2013), ibid. There are subtle differences between men and women in the relations between T and sexual behaviour and relationship orientation that, I suspect, may have to do with the effect of double standards on women's and men's willingness to report interest in casual sex.
55. Mazur, A., & Michalek, J. (1998). Marriage, divorce, and male testosterone. *Social Forces, 77*(1), 315–330. Quoted on p. 327.
56. Mazur & Michalek (1998), ibid. Quoted on p. 327.
57. Gettler, L., McDade, T., Feranil, A., & Kuzawa, C. (2011). Longitudinal evidence that fatherhood decreases testosterone in human males. *Proceedings of the National Academy of Sciences, 108*(39), 16194–16199.
58. Muller, M., Marlowe, F., Bugumba, R., & Ellison, P. (2009). Testosterone and paternal care in East African foragers and pastoralists. *Proceedings of the Royal Society B, 276*, 347–354.
59. Helpful reviews regarding testosterone and social status and testosterone and sexuality, respectively, are provided in Hamilton, L. D., Carré, J. M., Mehta, P. H., Olmstead, N., & Whitaker, J. D. (2015). Social neuroendocrinology of status: A review and future directions. *Adaptive Human Behavior and Physiology, 1*(2), 202 230; van Anders (2013), ibid.

60. van Anders (2013), ibid.
61. van Anders, S. M., Tolman, R. M., & Volling, B. L. (2012). Baby cries and nurturance affect testosterone in men. *Hormones and Behavior, 61*(1), 31–36.
62. A specific instance of a point made by van Anders (2013), ibid.
63. van Anders, S. M., Steiger, J., & Goldey, K. L. (2015). Effects of gendered behavior on testosterone in women and men. *Proceedings of the National Academy of Sciences, 112*(45), 13805–13810. Each actor did this twice: once in a stereotypically masculine way (for example, a cold expression and dominant posture), and once in a stereotypically feminine way (such as hesitant cadence and avoiding eye contact). This wasn't an important factor, indicating that T is linked with competitive behaviour per se, rather than masculinity.
64. van Anders et al. (2015), ibid. Quoted on p. 13808.
65. See van Anders & Watson (2006), ibid.
66. Oliveira, G. A., & Oliveira, R. F. (2014). Androgen responsiveness to competition in humans: The role of cognitive variables. *Neuroscience and Neuroeconomics, 3,* 19–32. Quoted on p. 21. Studies are summarized in Table 1, pp. 22–23.
67. One possible explanation for these inconsistencies is provided by the "dual-hormone hypothesis," according to which T levels interact with cortisol levels, such that T's positive effect on competitive behavior is blocked when cortisol levels are high. Mehta, P. H., & Josephs, R. A. (2010). Testosterone and cortisol jointly regulate dominance: Evidence for a dual-hormone hypothesis. *Hormones and Behavior, 58*(5), 898–906. For overview of the data, see Hamilton et al. (2015), ibid.
68. Oliveira & Oliveira (2014), ibid. Quoted on p. 23. For an empirical example, see Oliveira, G. A., Uceda, S., Oliveira, T., Fernandes, A., Garcia-Marques, T., & Oliveira, R. F. (2013). Threat perception and familiarity moderate the androgen response to competition in women. *Frontiers in Psychology, 4,* 389.
69. See the summary in Oliveira & Oliveira (2014), ibid.
70. Carré, J. M., Iselin, A.-M. R., Welker, K. M., Hariri, A. R., & Dodge, K. A. (2014). Testosterone reactivity to provocation mediates the effect of early intervention on aggressive behavior. *Psychological Science, 25*(5), 1140–1146. Quoted on p. 1140.
71. Carré et al. (2014), ibid. Quoted on p. 1144.
72. Cohen, D., Nisbett, R. E., Bowdle, B. F., & Schwarz, N. (1996). Insult, aggression, and the southern culture of honor: An "experimental ethnography." *Journal of Personality and Social Psychology, 70*(5), 945–960.
73. Cohen et al. (1996), ibid. Quoted on p. 957.
74. Herbert (2015), ibid. Quoted on p. 194.
75. Wade, L. (2013). The new science of sex difference. *Sociology Compass, 7*(4), 278–293. Quoted on p. 284.
76. For example, Bleier, R. (1984). *Science and gender: A critique of biology and its theories on women.* New York: Pergamon Press; Fausto-Sterling, A. (2012). *Sex/gender: Biology in a social world.* New York: Routledge.

77. Fuentes, A. (2012). *Race, monogamy, and other lies they told you: Busting myths about human nature.* Berkeley: University of California Press. Quoted on p. 16.
78. For example, Ridgeway, C. L. (2011). *Framed by gender: How gender inequality persists in the modern world.* Oxford, UK: Oxford University Press.
79. Liben, L. (2015). Probability values and human values in evaluating single-sex education. *Sex Roles, 72*(9–10), 401–426. Quoted on p. 415.

CHAPTER 7: THE MYTH OF THE LEHMAN SISTERS

1. Herbert, J. (2015). *Testosterone: Sex, power, and the will to win.* Oxford, UK: Oxford University Press. Quoted on pp. 116–118, reference removed.
2. Sunderland, R. (January 18, 2009). The real victims of this credit crunch? Women. *The Observer.* Retrieved from http://www.theguardian.com/lifeandstyle/2009/jan/18/women-credit-crunch-ruth-sunderland on January 15, 2015.
3. Prügl, E. (2012). "If Lehman Brothers had been Lehman Sisters . . .": Gender and myth in the aftermath of the financial crisis. *International Political Sociology, 6*(1), 21–35. Quoted on p. 21.
4. John Coates, interviewed in Adams, T. (June 18, 2011). Testosterone and high finance do not mix: So bring on the women. *The Guardian.* Retrieved from http://www.theguardian.com/world/2011/jun/19/neuroeconomics-women-city-financial-crash on February 20, 2014.
5. Kristof, N. (February 7, 2009). Mistresses of the universe. *New York Times.* Retrieved from http://www.nytimes.com/2009/02/08/opinion/08kristof.html?_r=0 on January 13, 2015.
6. Adams (2011), ibid.
7. Kristof (2009), ibid.
8. (July 13, 1902). Excluding women from brokers' offices; movement started in Wall Street to put an end to female speculating—reasons why brokers object to business of this kind—instances of woman's lack of business knowledge—why they are "bad losers." *New York Times.* Retrieved from http://query.nytimes.com/mem/archive-free/pdf?res=9502E0D9113BE733A25750C1A9619C946397D6CF on January 13, 2015.
9. *Time* cover on May 24, 2010. Cited in Nelson, J. (2013). Would women leaders have prevented the global financial crisis? Teaching critical thinking by questioning a question. *International Journal of Pluralism and Economics Education, 4*(2), 192–209.
10. The introduction of the term "big swinging dick" is credited to Lewis, M. (1989). *Liar's poker: Rising through the wreckage on Wall Street.* New York: Norton.
11. Croson, R., & Gneezy, U. (2009). Gender differences in preferences. *Journal of Economic Literature, 47*(2), 448–474. Quoted on p. 467.
12. Nelson, J. (2014a). The power of stereotyping and confirmation bias to overwhelm accurate assessment: The case of economics, gender, and risk aversion. *Journal of Economic Methodology, 21*(3), 211–231.

13. Nelson (2014a), ibid. See Table 1 on p. 216. Two studies obtained results encompassing greater *female* financial risk taking, with *d* ranging from −0.34 to null to 0.74, and from −0.25 to null to 0.49. In four further studies, no statistically significant differences were found. In five studies, results ranged from null to a low of *d*=0.37 and a high of *d*=0.85. In the final seven studies, the range of results spread from a low of *d*=0.06 to 0.17, to a high of *d*=0.55 to 1.13.
14. Nelson (2014a), ibid. Quoted on p. 212.
15. Stanley, T. D., & Doucouliagos, H. (2010). Picture this: A simple graph that reveals much ado about research. *Journal of Economic Surveys, 24*(1), 170–191.
16. More accurately, the *y*-axis plots "precision"; the inverse of standard error, which generally decreases with sample size.
17. Nelson (2014a), ibid. Quoted on p. 221.
18. Note this exercise treats "financial risk taking" as a single construct that, as Nelson notes, is an unexamined assumption.
19. Respectively: Hartog, J., Ferrer-i-Carbonell, A., & Jonker, N. (2002). Linking measured risk aversion to individual characteristics. *Kyklos, 55*(1), 3–26; Sunden, A. E., & Surette, B. J. (1998). Gender differences in the allocation of assets in retirement savings plans. *American Economic Review, 88*(2), 207–211; Barber, B. M., & Odean, T. (2001). Boys will be boys: Gender, overconfidence, and common stock investment. *Quarterly Journal of Economics, 116*(1), 261–292.
20. For instance, Hartog et al. (2002), ibid., found that risk aversion decreases with increased income and wealth.
21. Schubert, R., Brown, M., Gysler, M., & Brachinger, H. W. (1999). Financial decision-making: Are women really more risk-averse? *American Economic Review, 89*(2), 381–385. Interestingly, they also found that when abstract gambles were framed as losses (for example, Would you rather lose $30 for sure, or take a 50 per cent chance of losing $100?), women were significantly more risk taking than were men. But, again, this difference disappeared when the gambles were put into the less abstract context of insurance decisions. However, for contrasting findings, see Powell, M., & Ansic, D. (1997). Gender differences in risk behaviour in financial decision-making: An experimental analysis. *Journal of Economic Psychology, 18*(6), 605–628.
22. Vlaev, I., Kusev, P., Stewart, N., Aldrovandi, S., & Chater, N. (2010). Domain effects and financial risk attitudes. *Risk Analysis, 30*(9), 1374–1386. In this study, the researchers determined that the decisions fell into three kinds of financial decisions: positive (abstract "gain" gambles, pensions, and salary questions), positive and complex (mortgage and investment decisions), and negative (abstract "loss" gambles and insurance). There were no sex differences overall, or within each of these three groupings.
23. Henrich, J., & McElreath, R. (2002). Are peasants risk-averse decision makers? *Current Anthropology, 43*(1), 172–181.
24. When controlling for all other variables measured. Cameron, L., Erkal, N.,

Gangadharan, L., & Meng, X. (2013). Little emperors: Behavioral impacts of China's one-child policy. *Science, 339*(6122), 953–957.

25. Gneezy, U., Leonard, K. L., & List, J. A. (2009). Gender differences in competition: Evidence from a matrilineal and a patriarchal society. *Econometrica, 77*(5), 1637–1664. These studies used nontrivial stakes.
26. Gong, B., & Yang, C.-L. (2012). Gender differences in risk attitudes: Field experiments on the matrilineal Mosuo and the patriarchal Yi. *Journal of Economic Behavior and Organization, 83*(1), 59–65.
27. Cárdenas, J.-C., Dreber, A., von Essen, E., & Ranehill, E. (2012). Gender differences in competitiveness and risk taking: Comparing children in Colombia and Sweden. *Journal of Economic Behavior and Organization, 83*(1), 11–23.
28. Booth, A., & Nolen, P. (2012). Gender differences in risk behaviour: Does nurture matter? *Economic Journal, 122*(558), F56–F78; Booth, A., Cardona-Sosa, L., & Nolen, P. (2014). Gender differences in risk aversion: Do single-sex environments affect their development? *Journal of Economic Behavior and Organization, 99*, 126–154.
29. Sometimes economists define "risk" tasks as situations in which probabilities of pay-offs are known, and use "uncertainty" to describe situations in which the probabilities are not known. However, this convention is not followed here.
30. Cross, C. P., Copping, L. T., & Campbell, A. (2011). Sex differences in impulsivity: A meta-analysis. *Psychological Bulletin, 137*(1), 97–130. The effect size was $d=0.36$.
31. Cross et al. (2011), ibid. The effect size was $d=-0.34$. The authors suggest women are more likely to choose the high-risk packs because of greater punishment sensitivity. However, the high- and low-risk packs are equated overall for frequency of reward and punishment.
32. Holt, C. A., & Laury, S. K. (2002). Risk aversion and incentive effects. *American Economic Review, 92*(5), 1644–1655. See also Harbaugh, W., Krause, K., & Vesterlund, L. (2002). Risk attitudes of children and adults: Choices over small and large probability gains and losses. *Experimental Economics, 5*(1), 53–84. This study presented participants from ages 5 to 64 years with gambles involving "real and salient payoffs" (p. 55). The authors note "While many other researchers have found that men are less risk averse than women, with this protocol we find no evidence to support gender differences in risk behavior or in probability weighting, either in children or in adults." Quoted on p. 66, footnote removed. See also note 25.
33. Henrich & McElreath (2002), ibid. Quoted on pp. 175 and 175–176.
34. Akerlof, G. A., & Kranton, R. E. (2000). Economics and identity. *Quarterly Journal of Economics, 115*(3), 715–753.
35. Akerlof, G. A., & Kranton, R. E. (2010). *Identity economics: How our identities shape our work, wages, and well-being*. Princeton, NJ: Princeton University Press. Quoted on p. 10.
36. Akerlof & Kranton (2010), ibid. Quoted on p. 6.

37. For example, Nguyen, H., & Ryan, A. (2008). Does stereotype threat affect test performance of minorities and women? A meta-analysis of experimental evidence. *Journal of Applied Psychology, 93*(6), 1314–1334. For a more sceptical conclusion with respect to the magnitude of the stereotype threat effect, see Stoet, G., & Geary, D. C. (2012). Can stereotype threat explain the gender gap in mathematics performance and achievement? *Review of General Psychology, 16*(1), 93–102.
38. Carr, P. B., & Steele, C. M. (2010). Stereotype threat affects financial decision making. *Psychological Science, 21*(10), 1411–1416.
39. Brooks, A. W., Huang, L. Kearney, S. W., & Murray, F. E. (2014). Investors prefer entrepreneurial ventures pitched by attractive men. *Proceedings of the National Academy of Sciences, 111*(12), 4427–4431.
40. Gupta, V. K., Goktan, A. B., & Gunay, G. (2014). Gender differences in evaluation of new business opportunity: A stereotype threat perspective *Journal of Business Venturing, 29*, 273–288.
41. Gupta, V. K., Turban, D. B., Wasti, S. A., & Sikdar, A. (2009). The role of gender stereotypes in perceptions of entrepreneurs and intentions to become an entrepreneur. *Entrepreneurship Theory and Practice, 33*(2), 397–417.
42. Lemaster, P., & Strough, J. (2014). Beyond Mars and Venus: Understanding gender differences in financial risk tolerance. *Journal of Economic Psychology, 42*, 148–160; Meier-Pesti, K., & Penz, E. (2008). Sex or gender? Expanding the sex-based view by introducing masculinity and femininity as predictors of financial risk taking. *Journal of Economic Psychology, 29*(2), 180–196.
43. Twenge, J. (1997). Changes in masculine and feminine traits over time: A meta-analysis. *Sex Roles, 36*(5–6), 305–325.
44. Meier-Pesti & Penz (2008), ibid. This study primed masculinity and femininity by showing participants a picture of either a man in a business suit or a woman with a baby (or, in a control condition, a gender-neutral picture), and asking them to write about the scene, followed by similarly themed sentence completions.
45. Reinhard, M.-A., Stahlberg, D., & Messner, M. (2008). Failure as an asset for high-status persons: Relative group performance and attributed occupational success. *Journal of Experimental Social Psychology, 44*(3), 501–518; Reinhard, M.-A., Stahlberg, D., & Messner, M. (2009). When failing feels good: Relative prototypicality for a high-status group can counteract ego-threat after individual failure. *Journal of Experimental Social Psychology, 45*(4), 788–795.
46. Reinhard, M.-A., Schindler, S., & Stahlberg, D. (2013). The risk of male success and failure: How performance outcomes along with a high-status identity affect gender identification, risk behavior, and self-esteem. *Group Processes and Intergroup Relations, 17*(2), 200–220.
47. Weaver, J. R., Vandello, J. A., & Bosson, J. K. (2013). Intrepid, imprudent, or impetuous? The effects of gender threats on men's financial decisions. *Psychology of Men and Masculinity, 14*(2), 184–191. In the comparison condition, participants were asked to trial a power drill.

48. This second study used delay discounting as the dependent variable and found more impulsive behaviour in the masculinity threat condition.
49. Nelson, J. A. (2014b). Are women really more risk-averse than men? A re-analysis of the literature using expanded methods. *Journal of Economic Surveys, 29*(3), 566–585. Quoted on p. 576.
50. Beckmann, D., & Menkhoff, L. (2008). Will women be women? Analyzing the gender difference among financial experts. *Kyklos, 61*(3), 364–384.
51. Nelson (2014a), ibid. Quoted on p. 225.
52. Hönekopp, J., & Watson, S. (2010). Meta-analysis of digit ratio 2D:4D shows greater sex difference in the right hand. *American Journal of Human Biology, 22*, 619–630.
53. Voracek, M., Tran, U. S., & Dressler, S. G. (2010). Digit ratio (2D:4D) and sensation seeking: New data and meta-analysis. *Personality and Individual Differences, 48*(1), 72–77. Quoted on p. 76.
54. Herbert (2015), ibid. Quoted on p. 52.
55. Hönekopp, J., & Watson, S. (2011). Meta-analysis of the relationship between digit-ratio 2D:4D and aggression. *Personality and Individual Differences, 51*(4), 381–386. A small correlation was found for men only ($r=-.08$ for the left hand and $r=-.07$ for the right hand), but this reduced to a nonsignificant correlation for $r=-.03$ after correction for weak publication bias.
56. Voracek et al. (2010), ibid. The authors note the complexity of the biological system thought to underlie sensation seeking, as well as the many psychosocial factors known to influence it, and thus conclude that "Given these knowns, it appears unsurprising that rather simplistic approaches, such as studies only utilizing 2D:4D (a putative, not yet sufficiently validated marker of prenatal testosterone), are prone to be barren of results." Quoted on p. 76.
57. Vermeersch, H., T'Sjoen, G., Kaufman, J. M., & Vincke, J. (2008). 2D:4D, sex steroid hormones and human psychological sex differences. *Hormones and Behavior, 54*(2), 340–346.
58. Apicella, C., Carré, J., & Dreber, A. (2015). Testosterone and economic risk taking: A review. *Adaptive Human Behavior and Physiology, 1*(3), 358–385. Quoted on p. 369. Note that "risk taking" here was defined according to the economist's definition, thus referring specifically to lottery/gambling tasks. However, their subsequent review of 2D:4D findings for "risk-related constructs" reveals further inconsistencies.
59. Respectively, Apicella, C. L., Dreber, A., Campbell, B., Gray, P. B., Hoffman, M., & Little, A. C. (2008). Testosterone and financial risk preferences. *Evolution and Human Behavior, 29*(6), 384–390; Stanton, S. J., Mullette-Gillman, O. D. A., McLaurin, R. E., Kuhn, C. M., LaBar, K. S., Platt, M. L., et al. (2011). Low- and high-testosterone individuals exhibit decreased aversion to economic risk. *Psychological Science, 22*(4), 447–453; Schipper, B. C. (2014). Sex hormones and choice under risk. Working Papers, University of California, Department of Economics, No. 12, 7; Sapienza, P., Zingales, L., & Maestripieri, D. (2009). Gender differences in financial risk aversion and career

choices are affected by testosterone. *Proceedings of the National Academy of Sciences of the United States of America, 106*(36), 15268–15273; Doi, H., Nishitani, S., & Shinohara, K. (2015). Sex difference in the relationship between salivary testosterone and inter-temporal choice. *Hormones and Behavior, 69*, 50–58.

60. Stanton, S. J., Liening, S. H., & Schultheiss, O. C. (2011). Testosterone is positively associated with risk taking in the Iowa Gambling Task. *Hormones and Behavior, 59*(2), 252–256; Mehta, P. H., Welker, K. M., Zilioli, S., & Carré, J. M. (2015). Testosterone and cortisol jointly modulate risk-taking. *Psychoneuroendocrinology, 56*, 88–99.
61. Cueva, C., Roberts, R. E., Spencer, T., Rani, N., Tempest, M., Tobler, P. N., et al. (2015). Cortisol and testosterone increase financial risk taking and may destabilize markets. *Scientific Reports, 5*, 11206.
62. White, R. E., Thornhill, S., & Hampson, E. (2006). Entrepreneurs and evolutionary biology: The relationship between testosterone and new venture creation. *Organizational Behavior and Human Decision Processes, 100*(1), 21–34.
63. Sapienza et al. (2009), ibid. A correlation was seen when analyses were restricted to the lower testosterone range, as with the lottery task, but this assumes a nonlinear effect of testosterone, such that more risk taking is seen in individuals at intermediate levels, not high levels (that is, women with high testosterone and men with low testosterone, relative to their own sex).
64. Hewlett, S. (January 7, 2009). Too much testosterone on Wall Street? *HBR Blog Network*. Retrieved from http://blogs.hbr.org/2009/01/too-much-testosterone-on-wall/ on April 15, 2010.
65. Coates, J. M., & Herbert, J. (2008). Endogenous steroids and financial risk taking on a London trading floor. *Proceedings of the National Academy of Sciences of the United States of America, 105*(16), 6167–6172. It's assumed or inferred that this relationship between T and profits is due to increased risk taking.
66. Quoted in Solon, O. (July 13, 2012). "Testosterone is to blame for financial market crashes," says neuroscientist. *Wired*. Retrieved from http://www.wired.co.uk/news/archive/2012-07/13/testosterone-financial-crisis on January 13, 2015. Additionally, when things start to go badly in the market, large stress-triggered increases in cortisol are thought to adversely affect decision making.
67. Apicella, C. L., Dreber, A., & Mollerstrom, J. (2014). Salivary testosterone change following monetary wins and losses predicts future financial risk-taking. *Psychoneuroendocrinology, 39*, 58–64.
68. Leproult, R., & Van Cauter, E. (2011). Effect of 1 week of sleep restriction on testosterone levels in young healthy men. *JAMA, 305*(21), 2173–2174.
69. Of three testosterone manipulation studies with women, two failed to find any effects on risk taking on lottery tasks. Zethraeus, N., Kocoska-Maras, L., Ellingsen, T., von Schoultz, B., Hirschberg, A. L., & Johannesson, M.

(2009). A randomized trial of the effect of estrogen and testosterone on economic behavior. *Proceedings of the National Academy of Sciences of the United States of America, 106*(16), 6535–6538; Boksem, M. A. S., Mehta, P. H., Van den Bergh, B., van Son, V., Trautmann, S. T., Roelofs, K., et al. (2013). Testosterone inhibits trust but promotes reciprocity. *Psychological Science, 24*(11), 2306–2314. A small third study found that testosterone administration increased risk taking in the Iowa Gambling task, which typically finds slightly greater *female* risk-taking. van Honk, J., Schutter, D. J. L. G., Hermans, E. J., Putman, P., Tuiten, A., & Koppeschaar, H. (2004). Testosterone shifts the balance between sensitivity for punishment and reward in healthy young women. *Psychoneuroendocrinology, 29*(7), 937–943. The one study to manipulate testosterone in men found that those in the high testosterone group were more risk seeking in the Balloon Task, but not in the Iowa Gambling Task, or a third risk-taking task involving dice. Goudriaan, A. E., Lapauw, B., Ruige, J., Feyen, E., Kaufman, J. M., Brand, M., et al. (2010). The influence of high-normal testosterone levels on risk-taking in healthy males in a 1-week letrozole administration study. *Psychoneuroendocrinology, 35*(9), 1416–1421. More precisely, the intervention involved creating either high-normal testosterone levels and low-normal levels of estradiol, or low-normal testosterone levels and high-normal levels of estradiol.

70. Cueva et al. (2015), ibid.
71. See, for example, Solon (2012), ibid. Coates additionally suggests that since women are less susceptible to stress arising from financial competition, being more concerned with social competition, they will also be less susceptible than men to adverse effects on decision making from large increases in cortisol. However, the idea that a female trader would not find losing large sums of money stressful seems rather implausible.
72. For example, Ryan, M., Haslam, S., Hersby, M., Kulich, C., & Atkins, C. (2007). Opting out or pushed off the edge? The glass cliff and the precariousness of women's leadership positions. *Social and Personality Psychology Compass, 1*(1), 266–279.
73. Ibarra, H., Gratton, L. & Maznevski, M. (March 10, 2009). Claims that women are inherently more cautious are deeply troubling. *Financial Times*. Retrieved from http://www.ft.com/intl/cms/s/0/00829b22-0d14-11de-a555-0000779fd2ac.html#axzz3wbxCXczm on January 3, 2015.
74. Interview with *The Naked Scientists* (August 3, 2015). The truth behind testosterone. Transcript retrieved from http://www.thenakedscientists.com/HTML/interviews/interview/1001388/ on December 3, 2015.
75. According to Kristof (2009), ibid., the consensus at the World Economic Forum at Davos was that this would be the optimal bank.
76. Prügl (2012), ibid. Quoted on p. 22.
77. Nelson (2013), ibid. Quoted on pp. 205–206.
78. As described by U.S. senator Jim Bunning. Cited on p. 657 in McDowell, L. (2010). Capital culture revisited: Sex, testosterone and the city. *International Journal of Urban and Regional Research, 34*(3), 652–658.

PART THREE: FUTURE

1. Decent, T. (November 3, 2015). Melbourne Cup 2015: Winning jockey Michelle Payne hits back at doubters after making history on Prince of Penzance. *Sydney Morning Herald*. Retrieved from http://www.smh.com.au/sport/horseracing/melbourne-cup-2015-winning-jockey-michelle-payne-hits-back-at-doubters-after-making-history-on-prince-of-penzance-20151103-gkpouv.html on April 1, 2016.

CHAPTER 8: *VALE REX*

1. Women's Social and Political Union (1903).
2. See, for example, Auster, C., & Mansbach, C. (2012). The gender marketing of toys: An analysis of color and type of toy on the Disney store website. *Sex Roles, 67*(7–8), 375–388; Blakemore, J., & Centers, R. (2005). Characteristics of boys' and girls' toys. *Sex Roles, 53*(9/10), 619–633; Kahlenberg, S., & Hein, M. (2010). Progression on Nickelodeon? Gender-role stereotypes in toy commercials. *Sex Roles, 62*(11–12), 830–847.
3. Campaigns include Pink Stinks (*www.pinkstinks.co.uk*) and Let Toys Be Toys (*www.lettoysbetoys.org.uk*) in the United Kingdom and Play Unlimited in Australia (*www.playunlimited.org.au*). For critical comments by politicians, see BBC News (February 6, 2014), Aiming toys at just boys or girls hurts economy—minister. *BBC News*. Retrieved from http://www.bbc.com/news/uk-politics-26064302 on September 8, 2014; Paton, G. (January 16, 2014). "Gender specific toys 'put girls off' maths and science," says Education Minister. *The Telegraph*. Retrieved from http://www.telegraph.co.uk/education/educationnews/10578106/Gender-specific-toys-put-girls-off-maths-and-science.html on September 8, 2014. For critical commentary by psychologists, see Fine, C. (March 31, 2014). Biology doesn't justify gender divide for toys. *New Scientist*. Retrieved from http://www.newscientist.com/article/dn25306-biology-doesnt-justify-gender-divide-for-toys.html#.VA1TYfmSwjA on September 8, 2014; Hines, M. (July 12, 2013). There's no good reason to push pink toys on girls. *The Conversation*. Retrieved from http://theconversation.com/theres-no-good-reason-to-push-pink-toys-on-girls-15830 on September 10, 2013. For criticism by marketers see, for example, comments by Thinkbox chief executive Lindsey Clay, The Marketing Society Forum. (2014). Should all marketing to children be gender-neutral? *Marketing* (March 7). Retrieved from http://m.campaignlive.co.uk/article/1283685/marketing-children-gender-neutral on September 8, 2014.
4. Hoff Sommers, C. (December 6, 2012). You can give a boy a doll, but you can't make him play with it. *The Atlantic*. Retrieved from http://www.theatlantic.com/sexes/archive/2012/12/you-can-give-a-boy-a-doll-but-you-cant-make-him-play-with-it/265977/ on January 7, 2013.
5. The Marketing Society Forum (2014), ibid. Knox condemns marketing that reinforces prejudices or limits children's occupational aspirations—by

which he presumably means, for example, doctor and nurse sets labelled as "for boys" and "for girls," respectively.

6. Colarelli, S., & Dettmann, J. (2003). Intuitive evolutionary perspectives in marketing practices. *Psychology and Marketing, 20*(9), 837–865. Quoted on p. 858.
7. Saad, G. (2007). *The evolutionary bases of consumption*. Mahwah, NJ: Earlbaum. Quoted on p. 71.
8. Delingpole, J. (January 23, 2014). Why it's not sexist to say boys should never play with dolls. *Sunday Express*. Retrieved from http://www.express.co.uk/life-style/life/455465/Stop-making-our-children-neutral-let-boys-and-girls-play-with-gender-specific-toys on January 25, 2014.
9. Ireland, J. (December 2, 2014) "No gender 'December'": Greens senator calls for end to gender-based toys. *Sydney Morning Herald*. Retrieved from http://www.smh.com.au/federal-politics/political-news/no-gender-december-greens-senator-calls-for-end-to-genderbased-toys-20141202-11y4ro.html on April 27, 2015.
10. See (December 2, 2014), No gender-December—Don't let old-fashioned stereotypes limit children's festive fun. Retrieved from greens.org.au/node/6713 on January 3, 2015.
11. See Waters, L. (December 2, 2014). Let toys be toys. *The Hoopla*. Retrieved from http://thehoopla.com.au/let-toys-toys/ on January 3, 2015.
12. Wilson, L. (December 2, 2014). "Christmas shoppers should not buy gender based toys for kids," Greens say. *Daily Telegraph*. Retrieved from http://m.dailytelegraph.com.au/lifestyle/parenting/christmas-shoppers-should-not-buy-gender-based-toys-for-kids-greens-say/story-fniodobt-1227141319300 on August 3, 2015.
13. Ireland (2014), ibid.
14. Medhora, S. (December 2, 2014). No gender December: Abbott criticises bid to end gender stereotypes in toys. *The Guardian*. Retrieved from m/world/2014/dec/02/no-gender-december-abbott-criticises-bid-to-end-gender-stereotypes-in-toys August 3, 2014.
15. Tavris, C. (1992). *The mismeasure of woman: Why women are not the better sex, the inferior sex, or the opposite sex*. New York: Touchstone. Quoted on p. 212.
16. Pagel, M. (2012). *Wired for culture: Origins of the human social mind*. New York: Norton. Quoted on p. 4.
17. Wood, W., & Eagly, A. H. (2002). A cross-cultural analysis of the behavior of women and men: Implications for the origins of sex differences. *Psychological Bulletin, 128*(5), 699–727.
18. Fausto-Sterling, A. (2012). *Sex/gender: Biology in a social world*. New York: Routledge. Quoted on p. xiii.
19. This means that the developmental system is, in fact, an integral part of evolutionary processes, which act on individuals, not directly on individual genes. Jablonka, E., & Lamb, M. J. (2007). Précis of evolution in four dimensions. *Behavioral and Brain Sciences, 30*(4), 353–365.
20. Pagel, M. (2012). Adapted to culture. *Nature, 482*, 297–299. Quoted on p. 298.

21. Haun, D. B. M., Rekers, Y., & Tomasello, M. (2013). Children conform to the behavior of peers: Other great apes stick with what they know. *Psychological Science, 25*(12), 2160–2167.
22. Chudek, M., & Henrich, J. (2011). Culture-gene coevolution, norm-psychology and the emergence of human prosociality. *Trends in Cognitive Sciences, 15*(5), 218–226.
23. Sweet, E. V. (2013). *Boy builders and pink princesses* (Doctoral thesis). University of California Davis.
24. Connellan, J., Baron-Cohen, S., Wheelwright, S., Batki, A., & Ahluwalia, J. (2000). Sex differences in human neonatal social perception. *Infant Behavior and Development, 23*(1), 113–118.
25. See Levy, N. (2004). Book review: Understanding blindness. *Phenomenology and the Cognitive Sciences, 3*, 315–324; Nash, A., & Grossi, G. (2007). Picking Barbie's brain: Inherent sex differences in scientific ability? *Journal of Interdisciplinary Feminist Thought, 2*(1), 5.
26. Escudero, P., Robbins, R. A., & Johnson, S. P. (2013). Sex-related preferences for real and doll faces versus real and toy objects in young infants and adults. *Journal of Experimental Child Psychology, 116*(2), 367–379.
27. Zosuls, K., Ruble, D. N., & Tamis-LeMonda, C. S. (2014). Self-socialization of gender in African American, Dominican immigrant, and Mexican immigrant toddlers. *Child Development, 85*(6), 2202–2217.
28. Lamminmäki, A., Hines, M., Kuiri-Hänninen, T., Kilpeläinen, L., Dunkel, L., & Sankilampi, U. (2012). Testosterone measured in infancy predicts subsequent sex-typed behavior in boys and in girls. *Hormones and Behavior, 61*, 611–616.
29. Martin, C. L., & Ruble, D. N. (2004). Children's search for gender cues: Cognitive perspectives on gender development. *Current Directions in Psychological Science, 13*(2), 67–70.
30. LoBue, V., & DeLoache, J. S. (2011). Pretty in pink: The early development of gender-stereotyped colour preferences. *British Journal of Developmental Psychology, 29*(3), 656–667.
31. Shutts, K., Banaji, M. R., & Spelke, E. S. (2010). Social categories guide young children's preferences for novel objects. *Developmental Science, 13*(4), 599–610.
32. Hines, M., Pasterski, V., Spencer, D., Neufeld, S., Patalay, P., Hindmarsh, P. C., et al. (2016). Prenatal androgen exposure alters girls' responses to information indicating gender-appropriate behaviour. *Philosophical Transactions of the Royal Society B, 371*(1668). doi:http://dx.doi.org/10.1098/rstb.2015.0125
33. For example, Masters, J., Ford, M., Arend, R., Grotevant, H., & Clark, L. (1979). Modeling and labeling as integrated determinants of children's sex-typed imitative behavior. *Child Development, 50*, 364–371; Bradbard, M. R., & Endsley, R. C. (1983). The effects of sex-typed labeling on preschool children's information seeking and retention. *Sex Roles, 9*(2), 247–260.
34. Pasterski, V., Zucker, K.J., Hindmarsh, P. C., Hughes, I. A., Acerini, C., Spencer, D., et al. (2015). Increased cross-gender identification independent of gender role behavior in girls with congenital adrenal hyperplasia:

Results from a standardized assessment of 4- to 11-year-old children. *Archives of Sexual Behavior, 44*(5), 1363–1375.

35. Fine, C. (2010). *Delusions of gender: How our minds, society, and neurosexism create difference.* New York: Norton.
36. Jordan-Young, R. (2010). *Brain storm: The flaws in the science of sex differences.* Cambridge, MA: Harvard University Press; Jordan-Young, R. (2012). Hormones, context, and "brain gender": A review of evidence from congenital adrenal hyperplasia. *Social Science and Medicine, 74*(11), 1738–1744.
37. For example, Wolf, T. M. (1973). Effects of live modeled sex-inappropriate play behavior in a naturalistic setting. *Developmental Psychology, 9*(1), 120–123.
38. Wong, W., & Hines, M. (2015). Effects of gender color-coding on toddlers' gender-typical toy play. *Archives of Sexual Behavior, 44*(5), 1233–1242.
39. Wong & Hines (2015), ibid. Effect sizes for sex differences in play with a blue train and a pink doll at Time 2 were d=0.68 and d=–0.55, respectively. Effect sizes for sex differences in play with a pink train and a blue doll at Time 2 were d=0.26 and d=–0.21, respectively. A positive effect size corresponds to greater male interest.
40. Colyle, E. F., & Liben, L. S. (2016). Affecting girls' activity and job interests through play: The moderating roles of personal gender salience and game characteristics. *Child Development, 87*(2), 414–428.
41. Griffiths, P. E. (2002). What is innateness? *The Monist, 85*(1), 70–85.
42. For a recent articulation, see Gangestad, S. W., Haselton, M. G., & Buss, D. M. (2006). Evolutionary foundations of cultural variation: Evoked culture and mate preferences. *Psychological Inquiry, 17*(2), 75–95.
43. Griffiths (2002), ibid. Quoted on p. 74.
44. Dupré, J. (2001). *Human nature and the limits of science.* Oxford, UK: Oxford University Press. All quotations from p. 31.
45. Gottlieb, G. (2007). Probabilistic epigenesis. *Developmental Science, 10*(1), 1–11. Citing the work of Hood, K. (2005). Development as a dependent variable: Robert B. Cairns on the psychobiology of aggression. In D. M. Stoff & E. J. Susman (Eds.), *Developmental psychobiology of aggression* (pp. 225–251). New York: Cambridge University Press. These effects were also moderated by age and experience.
46. Rosenblatt, J. S. (1967). Nonhormonal basis of maternal behavior in the rat. *Science, 156*, 1512–1514.
47. Haslam, N. (2011). Genetic essentialism, neuroessentialism, and stigma: Comment on Dar-Nimrod & Heine (2011). *Psychological Bulletin, 137*(5), 819–824. Quoted on p. 819.
48. Brescoll, V., & LaFrance, M. (2004). The correlates and consequences of newspaper reports of research on sex differences. *Psychological Science, 15*(8), 515–520; Coleman, J., & Hong, Y.-Y. (2008). Beyond nature and nurture: The influence of lay gender theories on self-stereotyping. *Self and Identity, 7*(1), 34–53; Martin, C. L., & Parker, S. (1995). Folk theories about sex and race differences. *Personality and Social Psychology Bulletin, 21*(1), 45–57.
49. Skewes, L., Fine, C., & Haslam, N. (2015). *When boys will be boys, should*

women be women (and know their place)? Evidence from two nations on the relations between gender essentialism, gender bias, and backlash. Unpublished manuscript.

50. Gaunt, R. (2006). Biological essentialism, gender ideologies, and role attitudes: What determines parents' involvement in child care. *Sex Roles, 55*(7–8), 523–533.
51. Tinsley, C. H., Howell, T. M., & Amanatullah, E. T. (2015). Who should bring home the bacon? How deterministic views of gender constrain spousal wage preferences. *Organizational Behavior and Human Decision Processes, 126*, 37–48.
52. Dar-Nimrod, I., & Heine, S. (2006). Exposure to scientific theories affects women's math performance. *Science, 314*(5798), 435; Thoman, D., White, P., Yamawaki, N., & Koishi, H. (2008). Variations of gender-math stereotype content affect women's vulnerability to stereotype threat. *Sex Roles, 58*, 702–712.
53. Dar-Nimrod, I., Heine, S. J., Cheung, B. Y., & Schaller, M. (2011). Do scientific theories affect men's evaluations of sex crimes? *Aggressive Behavior, 37*(5), 440–449.
54. Keller, J. (2005). In genes we trust: The biological component of psychological essentialism and its relationship to mechanisms of motivated social cognition. *Journal of Personality and Social Psychology, 88*(4), 686–702; Morton, T., Haslam, S., & Hornsey, M. (2009). Theorizing gender in the face of social change: Is there anything essential about essentialism? *Journal of Personality and Social Psychology, 96*(3), 653–664.
55. Wood, W., & Eagly, A. (2012). Biosocial construction of sex differences and similarities in behavior. In J. Olson & M. Zanna (Eds.), *Advances in experimental social psychology* (Vol. 46, pp. 55–123). Burlington, MA: Academic Press.
56. England, P. (2010). The gender revolution: Uneven and stalled. *Gender and Society, 24*(2), 149–166.
57. Jordan-Young (2010), ibid. Quoted on p. 130.
58. Chiang, O. (January 7, 2011). Trojan: US market size for vibrators $1 billion twice the condom market size. *Forbes*. Retrieved from http://www.forbes.com/sites/oliverchiang/2011/01/07/trojan-us-market-size-for-vibrators-1-billon-twice-the-condom-market-size/ on January 8, 2015.
59. Jordan-Young (2010), ibid. Quoted on p. 113.
60. Meynell, L. (2008). The power and promise of developmental systems theory. *Les Ateliers de L'Éthique, 3*(2), 88–103. Quoted on p. 97, emphasis added.
61. McCormack, F. (June 24, 2015) How to prevent violence against women. Featured on *Big Ideas*. Retrieved from http://www.abc.net.au/radionational/programs/bigideas/fiona-mccormack-preventing-violence-against-women-in-australia/6552078 on June 27, 2015. See also Flood, M., & Pease, B. (2009). Factors influencing attitudes to violence against women. *Trauma, Violence, and Abuse, 10*(2), 125–142.
62. Ireland, J. (November 25, 2015). Greens get Senate inquiry to look into the

link between Barbies, toys and domestic violence. *Sydney Morning Herald*. Retrieved from http://www.smh.com.au/federal-politics/political-news/greens-link-barbies-trucks-and-childhood-toys-to-domestic-violence-in-call-for-gender-inquiry-20151124-gl716h.html on November 26, 2015.

63. Bigler, R., & Liben, L. (2007). Developmental intergroup theory: Explaining and reducing children's social stereotyping and prejudice. *Current Directions in Psychological Science, 16*(3), 162–166; Patterson, M., & Bigler, R. (2006). Preschool children's attention to environmental messages about groups: Social categorization and the origins of intergroup bias. *Child Development, 77*(4), 847–860.
64. Glick, P., Lameiras, M., Fiske, S. T., Eckes, T., Masser, B., Volpato, C., et al. (2004). Bad but bold: Ambivalent attitudes toward men predict gender inequality in 16 nations. *Journal of Personality and Social Psychology, 86*(5), 713–728.
65. For a comprehensive review, see Rudman, L., & Glick, P. (2008). *The social psychology of gender: How power and intimacy shape gender relations*. New York: Guilford Press.
66. Halim, M., Ruble, D., & Amodio, D. (2011). From pink frilly dresses to "one of the boys": A social-cognitive analysis of gender identity development and gender bias. *Social and Personality Psychology Compass, 5*(11), 933–949.
67. Cunningham, S. J., & Macrae, C. N. (2011). The colour of gender stereotyping. *British Journal of Psychology, 102*(3), 598–614. Quoted on p. 610.
68. See Roberts, Y. (September 13, 2015). Yet again men hold power. Why can't Labour change? *The Guardian*. Retrieved from http://www.theguardian.com/commentisfree/2015/sep/13/women-politics-power-labour-leadership-jeremy-corbyn on September 14, 2015.

INDEX